THE FALLACY OF MATERIALISM

ALSO BY THE AUTHOR

The Unity Principle: The Link between Science and Spirituality
The Nonlocal Universe: Why Science Validates the Spiritual Worldview
Reincarnation: Science of the Afterlife

Praise for *The Fallacy of Materialism*

"In *The Fallacy of Materialism*, Steven Richheimer, PhD, puts the ideology of materialism under the microscope and finds it severely wanting. In this sweeping analysis, Richheimer, a peerless scientist himself, presents solid reasons why consciousness is fundamental and immortality is our birthright. This magnificent work is sweeping in scope and inspirational in its message — a powerful, much-needed document for our time."

— **Larry Dossey, M.D.**, author of *One Mind: Why Our Mind Is Part of a Greater Consciousness and Why it Matters.*

"As a Stanford-trained chemist, author Steven Richheimer relied heavily on the strengths of the materialistic worldview for his daily work, and for good reason -- materialism is a highly accurate way to model some aspects of the physical world. But is it adequate to describe all of reality? In *The Fallacy of Materialism*, Dr. Richheimer clearly explains why the answer is a resounding no. He then offers an alternative view from the spiritual perspective and discusses why gaining a more comprehensive understanding of reality is more than just a nice idea. It's a critically important step to ensure the future health of ourselves and our civilization."

— **Dean Radin Ph.D.**, Chief Scientist, Institute of Noetic Sciences, author of *Real Magic,* and other books.

A thoughtful, surprising work on the nature of consciousness. Dr. Richheimer questions our assumptions about the basic structure of the universe in this work blending philosophy, spirituality, and physics.

What if the world is not made of atoms, as science teaches us, but of consciousness? Richheimer suggests this in his treatise on the ways modern people misunderstand the nature of reality. The materialism of the title refers not to consumerism of course, but to the concept of literal material: the upward causality theory of reality in which elementary particles form the basis of everything we experience. The alternative, which Richheimer calls spirituality, says just the opposite: "It postulates that consciousness is the 'ground substance' of creation and follows 'top down' causality since it hypothesizes that consciousness is transformed into cosmic mind and then into the material world." While spirituality seems almost by definition beyond the purview of science, Richheimer argues that recent developments in quantum mechanics and biology have suggested otherwise. He takes the reader through such mysterious territories as near-death experiences, the mind-body connection, animal instinct, and life after death. By eschewing dogmatism and approaching each topic with an open mind, the author hopes to challenge the reader's assumptions regarding how the universe really operates. Despite the specialized material, Richheimer's prose is accessible enough for the general reader to follow: "There is no question that psychedelic drugs can produce many of the same experiences described by mystics," he writes. "This has led many neuroscientists to label all mystical experiences as a product of abnormal brain chemistry." The author has a doctorate in chemistry, but the influence of New Age spiritualism suffuses his work. The book is dedicated to the spiritual guru Shrii Shrii Anandamurti—Richheimer's "spiritual preceptor and guide" and inspiration—and there's some talk of God. While the premise might alienate some readers, Richheimer approaches his arguments with honesty and curiosity and little preachiness. Those interested in the intersection of spirituality with contemporary hard science will find him to be a suitable guide: knowledgeable, nonjudgmental, and expansive.

— *Kirkus Reviews*

THE FALLACY OF MATERIALISM

HOW CONSCIOUSNESS CREATES THE
MATERIAL WORLD AND WHY IT MATTERS

Steven L. Richheimer, Ph.D.

InnerWorld Publications
San Germán, Puerto Rico
www.innerworldpublications.com

Copyright © 2021 by Steven L. Richheimer

Published in the United States by
InnerWorld Publications
PO Box 1613, San Germán, Puerto Rico, 00683

Library of Congress Control Number: 2021934772

ISBN: 9781881717836

Cover Design: Rodrigo Adolfo

All rights reserved. This book, or parts thereof, may not be reproduced in any form or by any means, electronic or mechanical, including photocopying, recording, or by any information storage or retrieval system, without permission of the publisher except for brief quotations.

ACKNOWLEDGEMENTS

My inspiration for writing this book came from my spiritual preceptor and guide Shrii Shrii Anandamurti, who I first had personal contact with in 1970 in Ranchi, India. He inspired me to diligently practice my meditation and become knowledgeable in spiritual ideology so that I might be better equipped to pass such knowledge onto others so that they may see the wisdom in entering the spiritual path of unity.

I am deeply indebted to Devashish Donald Acosta for his help in editing and laying out the manuscript for this book. I also wish to thank my wife and companion of 40 years, Jeanne, who inspired me to write this book and helped to edit it. I wish to thank Dada Maheshvarananda for his help in reviewing this manuscript for both the accuracy of the spiritual ideology presented and for other improvements to make it more readable.

DEDICATION

SHRII SHRII ANANDAMURTI

This book is dedicated to Prabhat Ranjan Sarkar (1922–1990) also known as Shrii Shrii Anandamurti or Baba (Father) to his disciples. He was a spiritual guru, philosopher, and composer. He founded Ananda Marga (the Path of Bliss) in 1955 as a spiritual and social service organization that continues to offer instruction in meditation and yoga. Sarkar was born in Jamalpur, India, and from a very early age he exhibited many extraordinary abilities, such as practicing meditation without learning it from a teacher; initiating much older persons in meditation; displaying great knowledge of languages, spiritual concepts, medicine, and various other topics, all gained without the help of teachers or books. He trained hundreds of yogic missionaries (acharyas and avadhutas) to spread his teachings of "self-realization and service to humanity" all over India and then throughout the world. The Ananda Marga organization eventually grew to become a large and multifaceted organization with members in over 130 countries, having different branches dedicated to the physical, psychic, and spiritual advancement of humanity. Anandamurti became known as a spiritual teacher, scientist, philosopher, neohumanist, social theorist, linguist, artist, and economist. He wrote over two hundred books on various subjects, such as history, spirituality, sociology, education, Tantra, yoga, medicine, ethics, psychology, humanities, linguistics, economics, ecology, farming, music, and literature. He gave several thousand discourses and composed over five thousand mystical songs known as *Prabhat Samgiita* (Songs of the New Dawn). He taught a system of yoga, meditation, and other spiritual practices to achieve all-around physical, mental, and spiritual development with the ultimate goal of spiritual union with cosmic consciousness.

CONTENTS

ACKNOWLEDGEMENTS	v
DEDICATION	vi
PREFACE	x
INTRODUCTION	1

PART ONE: THE PHYSICS OF REALITY — 5

1. CLASSICAL MATERIALISM	7
2. THE NEW MATERIALISM—SOME MODIFICATIONS NEEDED	14
3. THE WONDERFUL WORLD OF THE QUANTUM	29
4. INTERPRETATIONS OF QUANTUM MECHANICS	38
5. THE MYSTERIOUS DOMAIN OF THE WAVE FUNCTION	51
6. RELATIVITY THEORY	57

PART TWO: WHY MIND CANNOT BE REDUCED TO BRAIN — 67

7. HOW CAN REALITY BE EXPERIENCED?	71
8. THE MATERIALIST MODEL EQUATING MIND AND BRAIN	77
9. THE DIFFICULTIES IN EQUATING MIND AND BRAIN	86
10. MYSTICAL EXPERIENCES	96
11. OUT-OF-BODY AND NEAR-DEATH EXPERIENCES	103
12. PSI PHENOMENA	113
13. EVIDENCE FOR LIFE AFTER DEATH	129
14. FALSIFYING THE MATERIALIST WORLDVIEW	141

PART THREE: THE LOGIC OF THE SPIRITUAL WORLDVIEW — 149

15. HOW CONSCIOUSNESS IS TRANSFORMED INTO THE MATERIAL WORLD — 153
16. HOW SPIRITUAL IDEOLOGY EXPLAINS THE MYSTERIES OF QUANTUM PHYSICS AND RELATIVITY — 163
17. THE MIND-BODY CONNECTION — 173
18. UNDERSTANDING THE PARANORMAL — 182
19. LIFE AFTER DEATH — 187
20. EVOLUTION AND THE MYSTERY OF HOW LIFE BEGAN — 192
21. ANIMAL INSTINCTS AND HOMING ABILITIES — 200

PART FOUR: THE PROBLEM AND THE SOLUTION — 209

22. WHY SCIENTISTS FIND IT HARD TO REJECT MATERIALISM — 211
23. HOW MATERIALISM CONTRIBUTES TO SOCIETY'S PROBLEMS — 221
24. CHANGING SOCIETY FOR THE BETTER — 231
APPENDIX: A THEORY OF EVERYTHING — 241
BIBLIOGRAPHY — 260
ENDNOTES — 268
INDEX — 283

PREFACE

THIS BOOK IS ABOUT two opposing views of reality. One view can be termed *materialism* and postulates that the basic "stuff" of reality is matter. This worldview is a "bottom up" or upward causality theory of reality. Here elementary particles make up atoms, atoms combine to form molecules, molecules organize to make living cells with their all-important DNA, and some of these cells make the brain that produces subjective experiences including conscious awareness. In this scheme, mind and consciousness are epiphenomena of matter.

The alternative worldview about the nature of reality can be termed *idealism*—or as I prefer to call it *spirituality*.[1,2] It postulates that consciousness is the "ground substance" of creation and follows "top down" causality since it hypothesizes that consciousness is transformed into cosmic mind and then into the material world. Under the right conditions, living organisms may evolve on a planet and as a result unlock aspects of mind and consciousness that usually lie dormant in matter.

Theories and philosophies about reality normally come under the purview of metaphysics—not science—and are known as ontologies. Science is concerned with things that can be tested, measured, and quantified. It would certainly seem that questions about the nature of reality, existence of God, consciousness, etc. are not things that can be measured and should be left to philosophers—not scientists.

However, the discoveries of modern science—particularly those in quantum physics—have forced some scientists in the field to consider what these discoveries mean about reality. This is because of an "observation problem" that strongly suggests that consciousness may have a role in how physical reality manifests. In addition, there is considerable scientific evidence from the life sciences that indicates that mind cannot be reduced to brain (a postulate of materialism), and it is nonphysical as well as nonlocal.

My purpose in writing this book is to demonstrate that the evidence from science is clearly pointing to spirituality as the more credible theory of reality. Today it has become consensus among scientists to favor the materialist approach and certainly many intellectuals follow their lead. But this philosophy tends to degrade human existence to nothing more than the random dance of atoms that come together for an instant of cosmic time and then disappear forever. Every person is entitled to have their own ideas about reality and weigh the credibility of the various ontologies; however, they should not dismiss a body of evidence from one theory or another because they have already made up their mind that nature simply cannot work that way. Yet, this appears to be exactly what some of the strongest proponents of the materialist worldview do. They label all evidence that indicates mind is nonphysical, nonlocal, and survives death as pseudoscientific nonsense. They are convinced that the material universe is the ultimate reality and runs only according to known physical laws. They have no interest in investigating evidence of the paranormal since they feel that any evidence contrary to their belief system must be incorrect or fraudulent. This is known as dogmatism and is the same mind-set as that of religious fundamentalists who refuse to consider the evidence for evolution because they have already made up their mind that the biblical story of creation must be literally correct.

Obviously, you do not fall into this category since you have an interest in this book. All I ask is that you have an open mind—are an agnostic when it comes to whether there might be an alternate story about the nature of reality. My hope is to provide you with scientific evidence that will at the very least convince you that the materialist worldview has big holes in it. However, I do not believe that just poking holes in the reductionist-materialist view of reality is enough. If you are going to criticize someone's worldview, you must be able to propose an alternative worldview that is both logical and rational and provides an explanation for all the phenomena that you accuse the other worldview of ignoring, denying, or being wrong about. This is why I spend a good part of this book explaining how phenomena that might be considered anomalies of the materialist worldview are explained by spiritual ideology.

The spiritual ideology that I reference in this book is based on the numerous books, discourses, and private communications from my spiritual preceptor Shrii Shrii Anandamurti (aka Prabhat Ranjan Sarkar or Baba). For a short description of Anandamurti's remarkable life, please see the dedication of this book. His was a philosophy based on the ancient wisdom passed down from tantric and yogic sages, but was written for contemporary members of society. Of course, there are many variations to what I call spiritual ideology, but few are as comprehensive and all-encompassing as that of Anandamurti.

In this book, I have tried to avoid much of the jargon of both quantum physics and spirituality. Any description of quantum physics would be incomplete without introducing the concepts of the wave function and entanglement. At the same time, I assume the reader has a basic understanding of scientific principles such as the difference between a wave and a particle or an electron and a photon. Much of the basic jargon for spirituality is in Sanskrit. Sometimes the English language does not adequately express the more esoteric concepts of spiritual ideology, but I have endeavored to use the best English equivalents whenever possible. Finally, in order to avoid the religious and anthropomorphic connotations of the word God, I prefer to use such terms as cosmic consciousness, Creator, Cosmic Entity to describe the progenitor of the universe.

If I only increase your confidence that the spiritual worldview makes a lot of sense then I will have accomplished my purpose in writing this book. In the end, I believe the acceptance of a spiritual worldview requires a more expanded consciousness—i.e. a wider, more universal or spiritual outlook on life.

February 2021
Steamboat Springs, Colorado

INTRODUCTION

THE HISTORY OF THE human race indicates that human beings have always wondered about such things as whether their life has meaning; whether they may survive death in some new form; how the universe originated; whether there is a Creator; and how such a Creator might interact with them. Traditionally it was the world's great religions that attempted to answer these and other questions such as why there is evil, how and why one should behave morally, and how to obtain salvation.

The problem with most religions is that they fail to offer any practical way to know or experience God in this lifetime. Instead, they ask their followers to believe in doctrines based on faith, the word of their founder, or sacred texts. As a result, they do not provide any convincing evidence for the existence of God, nor for salvation in an afterlife. Hence, it is not surprising that organized religions have seen declining membership in this age of reason where people want proof, or at least an explanation of reality that is consistent with science. Furthermore, it is not surprising that science, which has been responsible for the many material benefits bestowed on humanity in the last few centuries, has to a large degree, filled the vacuum created by the crisis of faith that permeates society today.

Historically, religious beliefs often clashed with science. Scientific discoveries were sometimes dismissed by powerful religious institutions when the scientific findings contradicted Church doctrine. For example, Galileo was forced by the Catholic Church to renounce his findings that the Earth revolved around the sun or face imprisonment. Because of the tension between science and religion, science had to disconnect itself from anything religious or spiritual in order to proceed with its task of describing reality in a nonsuperstitious, nonmetaphysical, and fact-based manner. In fact, throughout history, science has discovered that many phenomena that were earlier thought to be mysterious or paranormal were actually explicable in natural terms. The spirit behind this movement in science is positive, but it has caused science to swing to the other extreme—materialism.

The use of the term materialism here does not refer to people's desire to attain happiness through the accumulation of material things or wealth. Rather the term is used to mean the philosophical theory of reality (ontology) that postulates that all natural phenomena including mind and consciousness can be explained in terms of material agencies. Materialism sometimes goes by other names such as scientific materialism, material realism, physicalism, reductionism, and naturalism. Not only are there different terminologies for this ontology, but there are also philosophical nuances to this theory of reality. However, they all accept the idea that physical matter along with its motions (energy) can explain all phenomena observed in the universe—there is no need to call on a Grand Designer to explain the origin and evolution of the universe into the one we witness today.

An alternate theory of reality can be called spirituality. It entails downward causality since it begins with consciousness and postulates that consciousness is transformed into mind and then into matter. In other words, it says that consciousness is the fundamental "stuff" from which the cruder aspects of reality are derived. This is a worldview that is consistent with most religions in that it postulates that reality and the universe emerged from something higher than matter—call it God—or the impersonal term I prefer—cosmic consciousness.

Materialists often dismiss this worldview as being dualistic. They accuse it of separating reality into two separate things—mind and matter. They say these are mutually exclusive and claim that it would be impossible for nonmaterial mind to interact with material matter. This is a false criticism since spiritual ideology is actually monistic—that is, everything can be reduced to consciousness.

Clearly, from a purely blank-slate perspective, both spirituality and materialism are equally plausible. In order to decide which vision of reality may be correct we need to consider the evidence that provides credence for each theory.

Part One of this book is entitled the physics of reality. We will look at the historical development of classical materialism, and how it had to be modified to fit the scientific evidence from the revolutionary theories of quantum physics and relativity. In order to understand what these two great theories indicate about the nature of reality, we will explore them in some depth.

We will explore several of the interpretations of quantum mechanics, which are designed to try to answer what may be the greatest question put to modern science—what does quantum mechanics imply about the fundamental nature of reality? We will learn that observation is required before any possible outcome described by quantum mechanics can become an actuality—i.e. emerge into physical reality.

No discussion of quantum mechanics would be complete without delving into the quantum wave function that is used to describe the possible states that a quantum system may take when observed. We will see how the domain of the wave function is one of wholeness and has properties identical to the cosmic mind postulated by spirituality.

We will look at Einstein's theories of relativity, and what they imply about the nature of reality. By proving that matter and energy, as well as time and space, are interchangeable, intimately connected, and have no reality independent of one another, these theories imply that the universe is best described as a singularity in which all the separate parts of the universe including all events—both past and future—are present in the wholeness of spacetime.

In Part Two of the book, we will consider how reality can be experienced. Since everything we know about physical reality comes to us through our brain, the question is whether we can even know if a physical reality exits that is separate from us.

Then we ask the question of whether mind can be reduced to brain. We will explore the considerable evidence that mind is nonphysical, nonlocal and cannot be reduced to brain. This evidence includes the difficulty that neuroscientists have with coming up with a good theory to explain consciousness. Secondly, there is the problem of explaining subjective experiences or qualia from a strictly physical basis. Then there is the problem of explaining self-awareness, memories, and intentionality. Neuroscientists cannot find any structure in the brain that generates self-awareness, nor can they come up with a theory of how intention is created by physical means or how memories are experienced in the third person and placed in time.

In addition, we will cover numerous phenomena in which a mental state produces a physiological change in the body. A few examples include the placebo effect, stigmata, hypnotic effects, skin writing, and control over autonomic functions.

Finally, we will explore the overwhelming evidence that several elements of human experience fail to be explained by materialism. These phenomena include mystical experiences, near-death and out-of-body experiences, ESP, and reincarnation memories. These experiences appear to be fundamental aspects of the spiritual nature of human beings. We will see that the data indicating that mind cannot be reduced to brain activity contradicts the theory of materialism that equates mind with brain, and says it is impossible for mind and consciousness to function without a physical body.

In Part Three of the book, we will take a plunge into the spiritual view of reality. To the Western mind, this ideology may at first glance seem strange and esoteric because it describes reality as holistic and our experience of a universe of physical objects that is separate from us as illusory. It describes reality as unfolding from unqualified cosmic consciousness. In this part of the book, we will learn about the detailed mechanism for how consciousness is transformed into the material world. In what can be dubbed the "creation cycle," consciousness is transformed into cosmic mind and then physical reality, under the influence of the trifunctional cosmic creative principle. We will learn how cosmic mind is both the creative and the unifying principle that is a subtler level of reality from which our experience of a relative reality emerges; and how it shapes the universe and sets the stage for the emergence of living organisms. In addition, we will look at how the existence of cosmic mind explains quantum phenomena, paranormal experiences, ESP, and evidence for a collective mind. Our analysis of the spiritual view of reality will show that it is logical and explains the anomalies or inconsistencies in the materialist worldview.

In Part Four of the book, we will learn why scientists find it difficult to reject materialism and how the adoption of the materialist worldview by a large portion of the public is a root cause of many of society's problems. Finally, we will consider how most of these problems would be solved by a change in people's attitude and understanding about the fundamental nature of reality as described by the spiritual worldview.

PART ONE

THE PHYSICS OF REALITY

1

CLASSICAL MATERIALISM

IN WESTERN CULTURE, PRIOR to the seventeenth century, the dominant way of viewing reality was through a religious lens that placed God in the central role of creator and master controller of the universe. The materialist worldview had not yet been well formulated. However, things began to change as a new scientific understanding of the world began to develop. In particular, French mathematician and philosopher René Descartes (1596-1650) saw a need to assign the objective sphere of matter to the domain of science and the subjective sphere of mind to the domain of religion.

By doing so Descartes was one of the most influential persons for laying the foundation for the modern concept of materialism. His was a dualistic philosophy that divided the world into an objective world that ran like a great machine according to physical laws described by science; and a world of mind that was nonmaterial and run according to God's rules. Hence, for Descartes mind and body were two entirely distinct substances—one was subject and the other was object. Even today, this Cartesian partition continues to influence human thinking. The idea that there is a separately existing objective reality out there, the "parts" of which are completely separate from us, is still deeply entrenched in the collective psyche of humankind and not likely to change anytime soon.

The one thing that Descartes could not adequately explain was how something nonmaterial (mind) could have an effect on something material (body). Nonetheless, his "substance dualism" meant that scientific investigations could be freed from the powerful influence of the Church. The ideas of Descartes were important in shaping classical scientific materialism, which is sometimes referred to as the Cartesian-Newtonian worldview.

Of course, the greatest contributor to the materialist worldview at the time was Isaac Newton (1642-1726). Newton's Laws of motion and gravity precisely described the motion of celestial bodies and physical objects. Newton postulated that space was infinite and unchanging—a three-dimensional fabric that always remained constant; and bodies were thought to move within it in precise, predictable ways. Time was also a constant within this rigid structure of space, and all clocks in the universe would tick at the same rate. This was the basis for the nineteenth-century concept of what is sometimes called a "clockwork universe."

While Newton's law of universal gravitation accurately described the orbits of celestial objects, it inferred an action at a distance. This greatly disturbed him because it conflicted with his other theories that were characterized by direct causation (vis-à-vis: action-reaction). Newton was also steadfastly religious, and he envisioned a universe that could move perfectly according to his laws of motion and gravity, but nonetheless he argued that a Supreme Being or Intelligence must have set it up. In other words, a universe that functioned like a great clock must have a Master Clockmaker who created it.

The other great contributor to classical materialism was the brilliant French mathematician and astronomer Pierre-Simon Laplace (1749-1827). Unlike Newton, Laplace leaned toward atheism. He had an unparalleled understanding of Newtonian physics and took it to a new level of applicability by improving the mathematics (calculus), and introducing the concept of field theory. Field theory is the idea that a force such as gravity permeates space, and is able to act at a distance—its force diminishing with increased distance. This solved Newton's action-at-a-distance problem.

At about the same time James Maxwell developed his equations describing electricity and magnetism and Charles Darwin proposed his theory of evolution that provided a reasonable explanation for the evolution of complex living organisms. Now for the first time scientists began to see the universe as a finely tuned mechanistic system—there was no need to call on a Creator to explain the working of the universe and origin of complex living organisms. Scientists at the time believed that they were on the verge of explaining everything in terms of matter, energy, and physical forces.

The idea that physics had "wrapped up" its work in explaining the universe was expressed by several notable scientists at the time. For example, Albert Michelson stated in 1894, "The more important fundamental laws and facts of physical science have all been discovered." Astronomer Simon Newcomb in 1888 wrote, "We are probably nearing the limit of all we can know about astronomy." Interestingly, Newcomb is also famous for saying that it would be impossible to build a "flying machine." He argued that the construction of such an aerial vehicle would require the discovery of some new force. Then we have Sir William Thompson (Lord Kelvin), the distinguished English physicist, who helped formulate the first and second laws of thermodynamics, proclaiming in 1900 that physics was complete except for "two clouds." As it turned out, these clouds led Max Planck to the discovery of quantum mechanics and Einstein to propose relativity theory.[3] Physics professors around the turn of the century were even advising their students to consider another career path since all the important matters in physics had already been discovered.

The field of science is littered with such predictions and claims that were later proven false. As we will see, the postulates of the Cartesian-Newtonian worldview were "blown up" by the discoveries of quantum physics and relativity that occurred in the early part of the twentieth century. Physicists were forced to amend the classical theory of materialism to fit with the new model of the universe described by these two great theories.

The Postulates

Every theory relies on postulates or assumptions and then builds its case based on these—the "foundation blocks" for that theory. Classical materialism is no different. It assumes that matter and its motivating force (energy) are ultimate reality. Everything is built from matter and there is no subtler level of reality than matter. Living organisms, mind, and consciousness originate from matter, and there is no scope or need for a Creator of the universe. Materialism may be described as a "bottom up" ontology, because everything is built up from matter (atoms or their subatomic building blocks). Complex, self-replicating biomolecules originate due to chemical transformations that give rise to simple single-celled living organisms. These simple life forms experience environmental and competitive pressures, undergo natural variations, and with increased survival of beneficial traits, evolve into increasingly complex life forms in which mind and consciousness emerge.

One of the postulates of this theory is that one should be able to predict with absolute certainty the state of a system in the next instant if one knows the current conditions of the system and applies the laws of physics. In other words, the future can be predicted with certainty if precise knowledge of the present is known. This is called *causal determinism*, and it was first described by Laplace.

In addition to determinism, a second key postulate of materialism is *locality of matter-energy interactions*. Objects, no matter what size or mass, are governed only by local force fields that diminish with distance. One should be able to pinpoint the location of any particle in three-dimensional space using the Cartesian system of xyz variables. Additionally, there is no theoretical limit on the speed of an object, nor would such motion have any effect on the flow of time, which is a constant throughout the universe.

Since the theory claims that mind and consciousness originate from matter, there is the requirement that mind and consciousness have no existence outside the brain/nervous system (epiphenomenalism/locality of mind). According to this postulate, physical objects exist independent of one another and there is no scope for mind, which is created by the brain to have an effect on any physical entity except the body. This postulate of materialism can be called *strong objectivity*.

It should be immediately obvious to anyone acquainted with modern science that several of these postulates of classical materialism are now known to be false. This is primarily based on the discoveries of quantum physics and relativity in the last century. This caused scientists that were wedded to the materialist worldview to modify the theory. These modifications are discussed in the next chapter, but do not affect any of the basic implications of the theory.

The Creed of Scientific Materialism

By equating mind with brain activity, materialism states that we are nothing but extremely complex biochemical machines and everything we are, do, and feel can be explained causally by physical processes. In other words, all mental phenomena are created by the interaction of tiny bits of matter and movement of electrical impulses in accordance with the known laws of physics. The consequences of this worldview are outlined below. They constitute a worldview that can be called the "Creed of Scientific Materialism."

- Our experience of reality comes to us through our sense organs and we are assured of the reality of this domain because of the general agreement about its nature among human beings.
- Such things as free will, intention, and self are entirely the creation of electrochemical processes taking place in the brain and are therefore illusory. This also goes for such things as love, awe, altruism, creativity, magnanimity, humility, serenity, gentleness, mercy, intuition, conscience, serenity, and spiritual ecstasy.
- Every thought, emotion, and memory we ever had or will have is directly tied to neural activity. Even consciousness is the direct result of electrical impulses occurring in our vast network of brain cells.
- There is no God. We have no destiny other than to live in this body but for a miniscule slice of cosmological time. There is no meaning or purpose to life. Any effect we might have on those near us or on society will undoubtedly be minimal and all but forgotten after a few generations following our death. Any ideas about God, prophets, seers, enlightened humans or other nonphysical beings are superstitions and delusions. Supernatural forces or powers cannot exist. Any stories of persons demonstrating such powers are myths.
- Since life has no higher purpose or meaning other than the basic biological drives we need to live by the doctrine that what gives us pleasure and happiness is good, and what gives us pain, and unhappiness is bad. Our friends bring us happiness and those that prevent us from such happiness must be considered our enemies. There is nothing like spirit connecting all of humankind. Any talk of others outside our small sphere of family being our brothers and sisters is nonsense.
- All our judgments of right and wrong, ideas of morality, and values are subjective—not handed down from some "higher authority," but the product of evolutionary advantage.
- There is no karmic debt to pay for our antisocial actions, just the possibility of being punished by society. Similarly, our good actions do not create good reactions that we will reap in the future.
- All transcendental, mystical, and out-of-body experiences are caused by "abnormal" or drug-induced psychological states. Reports of lucid consciousness continuing during cardiac arrest or while the brain is shut down because of a lack of blood flow

(near-death experiences), must be due to physiological causes—for example, the massive release of stress chemicals. The brain generates consciousness and it is impossible for consciousness to either leave the body or continue for more than a few seconds after the brain shuts down.
- All paranormal and psychic phenomena characterized as extrasensory experiences (ESP) are impossible. Any evidence to the contrary is either bad science, wishful thinking, or fraudulent. Such phenomena would clearly violate the known laws of physics and are impossible because mind is localized—generated by the brain.
- Life on Earth began as the result of random chemical permutations that took place in the sea, which was rich in organic compounds with energy supplied by the sun or hydrothermal vents.
- Evolution of complex systems and species is unplanned and unguided. Whether genetic information is passed on to future generations depends on how well an organism adapts to its environment and random chance. Physical traits that increase an organism's survival tend to be preserved and those that are neutral or decrease its chance of survival die out (selection of the fittest).
- Evolution is not teleological. Complex systems such as the immune system that might need several mutations before any advantage is conferred to the organism nonetheless evolve in a stepwise manner—even if the probability of this occurring by chance is very low.
- Mind and consciousness "emerge" in living organisms after a certain level of complexity develops and are a result of brain activity. There is no good theory of where they come from, but apparently, they exist in a potential form in matter.
- All psychological states and qualia (phenomenal experiences and emotions) correspond to different neuronal patterns in the brain.
- With dissolution of the body and brain following death, there is no possibility that mind, consciousness, memories, and elements of personality survive. Any evidence of children or adults having veridical memories of living in a previous body (reincarnation) must be false—coincidental, exaggerated, or fraudulent anecdotes.

To summarize, we may conclude that materialism maintains that we are part of a physical world, patterns of atoms that behave only according to the laws of physics. We should be considered as nothing more than a complex collection of matter moving according to physical laws—merely sophisticated robots. Mental phenomena are caused by neurophysiological processes in the brain. In other words, mind is brain. Hence, any evidence suggesting that mind and consciousness are anything other than generated by brain activity is faulty science (pseudoscience) or concocted. A Creator is unnecessary for explaining our existence and, in fact, does not exist.

2

THE NEW MATERIALISM—SOME MODIFICATIONS NEEDED

FROM A SCIENTIFIC POINT of view, a theory must be rejected or modified if its predictions fail to account for observed physical phenomena. In the first part of the twentieth century, the scientific evidence supporting the revolutionary theories of quantum mechanics and relativity forced scientists to formulate a new model for physical reality. Although these two theories could not be reconciled because of some fundamental differences in how they describe reality,[4] several of the postulates of classical materialism had to be thrown out or modified in order to come up with an ontology that was consistent with the new body of scientific evidence that supported these two great theories.

What are Quanta Anyway?

A quantum refers to a discrete amount of energy. Instead of the emission of energy being continuous from a source such as a heated piece of metal, it is observed to exist as discrete packets of energy. These energy packets are called quanta—a term first coined by Max Planck in 1900 to describe the quantized nature of electromagnetic radiation from a hot blackbody.[5] Planck postulated that electromagnetic energy could be emitted only in discrete packets (called photons) equal to the frequency of the radiation times a constant, which is known as Planck's constant (h).

This discovery represented the birth of quantum theory and the beginning of the end for classical physics. Finally, it follows that if energy can have particle-like properties then particles—which by definition have mass and are considered discrete objects—can have wave-like properties. With the discovery by Einstein of a mathematical relationship of mass and energy ($E=mc^2$), physicists now consider tiny particles such as electrons, protons, neutrons, quarks, and even atoms and molecules as "quantum objects." These have both particle and wave characteristics and therefore are considered "quantum particles."

Separability and the Quantum Wave Function

Materialism postulates that there is a universe out there consisting of physical objects that are separate from us. This is quite reasonable since everyone seems to experience the same reality, and we are not able to create, change, or move physical objects with our minds. However, separability between observer and observed is not seen in the quantum realm. For example, how we design an experiment to observe a quantum particle will influence whether it behaves as a particle or a wave. Countless experiments have shown that quantum events are influenced by the observer. In other words, whenever we observe a quantum system with any of our measuring devices we change the system. At best, one could say that the objectivity of quantum mechanics is weak in that the outcome of the observation of a quantum event is not dependent on *who* makes the observation.

Since we cannot know the state of a particle before it is measured, quantum theory concludes it must be a superposition of all possible states. A superposition is the collection of all the possible states a quantum particle (or system) can assume when observed. The real utility of quantum mechanics is that these possible states can be described mathematically by what is called a "quantum wave function." Not unlike water or vibrating violin strings, the quantum waves oscillate and have crests and troughs; and these high and low spots in the wave function correlate with the probability or likelihood that any of the possible states in the superposition will pop into reality when the system is observed.[6]

In addition, there is a way to describe how the wave function evolves with time. The equation for this is called the Schrödinger equation because it was first developed by Nobel Prize-winning physicist Erwin Schrödinger.

An example of a wave function is the one that describes the hydrogen atom. The wave equation for the electron in this simplest of atoms describes all the possible locations and energy levels in which the electron could be found when we observe it. Like any wave, its amplitude varies corresponding to the various possible locations. Finally, if the wave's amplitude at any location is squared one obtains the probability of observing the electron at that location. The only requirement is that all the probabilities add up to one since the electron must be found somewhere. It turns out that the solution to this particular wave function indicates that only certain locations corresponding to different energy levels are allowed. These levels are referred to as shells or orbits. In other words, the energy levels are "quantized" not continuous. Furthermore, it turns out that when the electron moves from one shell to another that it does so without passing through intermediate energy levels. This is what is known as a "quantum leap." And this magical ability to jump instantaneously from one location to another without traversing a path between the two locations is a universal quantum phenomenon. It is as though quanta can transcend the limitations of space and time and flicker into and out of existence with no known physical process connecting the two states of their being.

If we observe or measure the electron, our observation causes it to assume only *one* of the many possible states described by the wave function. In other words, observation causes the wave function, which can theoretically incorporate an almost infinite number of possibilities, to *collapse into a single state*. This process is termed the "collapse of the wave function." Not all physicists like using this term because it implies that only one possible outcome is generated by observation; but they all agree that the superposition is broken by observation and the system evolves according to the Schrödinger wave equation.

The quantum wave function is a powerful tool used by physicists to investigate quantum phenomena because it provides both the range of possibilities or potential states for a quantum system as well as the likelihood that any particular state will manifest in physical reality upon observation. According to quantum mechanics, it is the wave function that is *real*, not the particle that we observe. Additionally, everything in the universe including the universe itself must have a wave function since there cannot be any fundamental difference between a microscopic entity such as an electron and a macroscopic entity such as a baseball—where would you draw the line?

Quantum mechanics is the deepest and most comprehensive view of reality that science has discovered. At the heart of this incredible discovery about nature is the wave function. Yet, this wave function details possibilities, and describes a hidden domain from which physical reality emerges. Hence, it speaks of a level of reality that is subtler than physical reality. It is no longer possible for scientists or philosophers to argue that matter is the most fundamental level of reality or that reality consists of separately existing physical objects. The hypothesis of materialism that the world consists of independently acting particles has been shown to be false by quantum physics forcing advocates of this worldview to modify their ontology.

The Locality Problem

You might recall that according to classical materialism, one should be able to locate precisely any particle of matter in space using sensitive instruments and then predict the particle's exact path or position in the near future. One simply needs to know all the information about the forces influencing it.

However, this is not possible for quantum particles because even with the best instruments theoretically possible there is always a small degree of uncertainty in the measurement of such things as speed and position. According to the well-established evidence of quantum mechanics, the better the speed (or momentum) of a particle is known, the less we can know about its position. And when we zero in on a tiny particle's position, we discover we know less about its speed. This problem is best described by what physicists call the Heisenberg uncertainty principle, which describes the minimum uncertainty that any quantum system can have, and it is a direct result of the fact that quantum particles are not discrete, but best described as wave-like.

Not only is it impossible to know with certainty the speed and position of a quantum particle, where it will be found when observed is governed by probability rather than certainty. For example, we might expect to find an electron at a certain location on a cathode ray tube (an instrument that can light up like an old TV screen when bombarded by electrons). Although the chances are good we will find the electron right where we expect it, there is a very small, but nonetheless, finite probability of finding it in another galaxy.

Although these violations of locality can be considered rather small and not applicable to large objects like baseballs, there is a much more grievous violation to the materialist tenet of locality—*entanglement*. Entanglement is the phenomenon in which you have two (or more) quantum particles or systems that do not behave as separate entities. Most often the quanta are born together in a process that creates them, and they remain correlated or connected no matter how far apart they are; and when one of the quanta is observed, its quantum state is immediately communicated to its twin. In the language of quantum mechanics, we say that the two quanta *share a single wave function*, and when one is observed the wave function for the pair collapses and both quanta take on definite properties. It does not matter how far apart they are and the effect is instantaneous.

This prediction of quantum mechanics was so bizarre that most physicists at the time, including Albert Einstein, thought that quantum mechanics must be a flawed or incomplete theory. However, subsequent experiments conducted since the 1980s have proven beyond any doubt that entanglement is a reality. We now know that information about the quantum state of one quantum is communicated to other entangled quanta instantaneously, and that no signal is involved. In addition, the strength of the connection does not fade with increased distance. Indeed, today entanglement is taken for granted by physicists and being investigated and used in ways that will surely revolutionize the fields of electronics and communication.[7]

The connection that exists between distant but entangled quanta is now an established quality of nature. It is proof that nonlocality exists in the physical realm.[8] It disproves another of the key postulates of classical materialism—locality—forcing advocates to this worldview to modify the theory.

Determinism Is Violated

Quantum mechanics has also refuted the materialist principle of causal determinism. Strong causal determinism assumes that if all the initial conditions of a system are known then any change in the system caused by an outside agent (whose condition is also known) can be predicted precisely. The problem was that for the new science of quantum mechanics, there is always an element of randomness and unpredictability inherent in the system. For example, radioactivity is both random and unpredictable. No amount of foreknowledge can predict how and when an unstable atom will decay. In addition, all quantum phenomena have a small degree of uncertainty (recall the Heisenberg uncertainty principle), which means that it is never possible to describe the system exactly. According to quantum physics, causal determinism does not apply at the fundamental level of matter and energy—only probabilities.

Does Mind Affect Matter?

Classical materialism claims that mind and consciousness originate from matter. Therefore, it would be impossible for mind and consciousness to exist outside the material brain and nervous system (locality of mind). According to this postulate, mind cannot affect a physical object other than the body. If mind were to be considered separate from the brain, it would be a nonphysical entity, and no mechanism exists by which something nonphysical could affect a physical object such as a brain neuron. In addition, any such interaction would violate the law of conservation of energy.

However, there are problems with both of these arguments. First, there is the "observation problem" posed by quantum mechanics. This indicates that wave function collapse only occurs after the system is observed. Hence, mind/consciousness appears to be directly involved in how reality emerges from the superposition described by the wave function. In other words, the wave nature of a quantum particle is spread throughout the entire universe until it becomes localized by some sort of observation or measurement. The collapse of the wave function could be initiated by an instrument such as an oscilloscope run by a computer that was set up by a scientist a week before. It could also be caused by a photographic plate or the human eye (assuming it had the required resolution). Quantum mechanics does not differentiate between any of these forms of observation. All would collapse the wave function. The one thing that is common for all of these is that a human is required to perform the observation or set up the measuring apparatus. Therefore, for every act of observation, some form of mind or consciousness is involved and must be thought of as having caused the wave function collapse. This is the crux of what has been termed the observation or measurement problem of quantum mechanics. Many have tried to explain this phenomenon, and it is largely responsible for all the commotion regarding what quantum mechanics is telling us about the nature of reality. It has even spawned the new field of study that is termed the "interpretations of quantum mechanics." This is a topic covered in Chapter 4.

The second falsehood is the assumption that something nonphysical (mind) cannot influence a physical object (e.g. the brain) because in doing so it would violate the law of conservation of energy. However, if mind is nonphysical and can function separately from the brain, it could still be capable of influencing the brain on a quantum level. This would not require the mind to expend any energy in the process, since it is well known that the simple act of observation causes wave function collapse, which could result in a particular probability event to take place in a physical organ (e.g. neurons in the brain). No energy is required, only intention or directed consciousness. Hence, the conservation of energy as normally understood does not apply under these circumstances.

Big Bang, Entropy, and the Arrow of Time

Cosmologists believe that time and space began in a gigantic expansion from a dimensionless point or singularity in what is commonly called the "Big Bang." The principal reason for this theory is that the universe is known to be expanding. Edwin Hubble was the first astronomer to discover that distant galaxies were all moving away from Earth and from each other at a rate proportional to the distance separating them. This is called Hubble's law and the rate of expansion is called the "expansion factor" or "scale factor." Hence, if a galaxy is one hundred million light years distant from us, it will be observed to be receding from us at half the velocity of a galaxy that is two hundred million light years distant. Subsequent studies have shown that this rate appears to be increasing with time. By turning back the clock on this expansion of space, cosmologists estimate that our universe is 13.8 billion years old.

The Big Bang model says that the expansion of the universe is due to the expansion of space itself. If this were not so then distant galaxies would be moving away from us at a speed greater than light—which is not possible. Another consequence of this type of expansion is that the universe, which began as a point, has no center. No matter where you stand in the universe, everything is observed to be receding from you at the same relative rate. Hence, from a scientific point of view the religious doctrine that we are at the center of the universe is actually correct, but nonetheless meaningless since every point in the universe could be considered the center.

The Big Bang model of cosmogenesis does have a few anomalies that are solved by adding a new process called "cosmic inflation." This occurred immediately after the birth of the universe.[9] According to this theory, the universe initially expanded very rapidly—much faster than the speed of light. Since it was space that expanded, this does not contradict the theory of relativity, since the speed of light was not affected. At the end of inflation, the universe decayed into the normal expansion observed today.

Following inflation the universe was still too hot for ordinary matter to exist, in less than a few million years the universe cooled sufficiently to allow the formation of subatomic particles, and then atoms—mostly hydrogen in addition to small amounts of helium and lithium.

Cosmologists describe the initial dimensionless point of creation as one of *minimum entropy*. Entropy is a thermodynamic property that is an important component of the second law of thermodynamics. Entropy is needed to explain certain reactions that occur spontaneously without the expenditure of normal energy—like heat. For example, consider two containers; one filled with oxygen and the other with nitrogen. Now connect the containers by a small tube. Gradually oxygen molecules from container A will pass into container B while nitrogen molecules in B will pass into A. Eventually, the two gases will become completely mixed. The driver for this mixing is called entropy. The system goes from a highly ordered state (two pure gases) to a less ordered state (mixed gases). Hence, entropy is the hidden energy causing greater disorder or randomness, and it is ever increasing in the universe. It is behind the formation of stars, galaxies, planets, and living systems.

Now it might be fair to ask the question of whether living organisms violate the rule that entropy is always on the increase. Unlike your two-year-old who seems to contribute to global randomness, many human endeavors involve constructing well-ordered things starting from disordered raw materials. However, living organisms do not actually violate the second law of thermodynamics with its requirement that increasing entropy drives change in most everything. When the results of the metabolism of highly ordered food molecules are taken into account, the net effect is still an increase in entropy. Ultimately, all the energies utilized by living organisms can be traced back to the energy of the sun, which by converting mostly hydrogen into heavier elements and lots of energy, contributes to an increase of entropy in the universe.

One way to understand how increasing entropy or disorderliness occurs in the universe over time is in terms of probable outcomes. For example, consider a new deck of cards, which comes with all four suits ordered from ace to king plus two jokers. If you throw the deck of cards into the air, it will probably not return to its original order. This is because there are millions of possible ways the cards could fall in a disordered fashion and only one way they could return to their original state. Hence, in all likelihood the cards will become disordered.

Entropy also provides the arrow of time for the universe. All the laws of physics work fine in both temporal directions—time flowing forward or backward. Our experience of the flow of time is tied to our experience of increasing randomness. We drop a glass on the floor and it shatters into a thousand pieces. We never witness crumbled brick coming together to form a whole brick. We identify the past with a whole egg and the future with its shell being cracked and its contents scrambled. The arrow of time for the universe exists because of the Big Bang, which was maximally ordered, and the fact that disorder constantly increases as we look forward in time.

Prior to the Big Bang, neither space nor time existed. Therefore, it is meaningless for scientists to hypothesize about what came before the Big Bang. Both space and time began at the time of the Big Bang and since Einstein's theories of relativity demonstrate that these two cannot be considered separate—but form a four-dimensional continuum—the correct nomenclature is to say that spacetime was created and expanded following the Big Bang.

Although we experience the linear flow of time and it plays an important role in the physical sciences such as chemistry, geology, astronomy, physics, and biology, the linear flow of time is a classical concept—in both quantum and relativistic mechanics—the directional flow of time is meaningless.

The Fine-tuning Problem

Newton envisioned a universe that was created by a "Master Watchmaker" (God). God's design included all the physical laws and materials needed for the entire enterprise to function perfectly and allow for conditions here on Earth to be suitable for life. However, no such concept of a "Grand Designer" is consistent with the materialist worldview. Scientists now recognize that the cosmological constant (energy of vacuum space); the total mass-energy of the universe; the fundamental forces; the physical laws; the particle masses; and the constants—all seem to be exquisitely fine-tuned to allow for the existence of stars, planets, water, and ultimately for the emergence of complex living organisms.

In order for the universe to be long-lived, there had to be a very fine balance between the total mass-energy of the universe and the force of gravity. If the universe following the Big Bang had only a tiny bit more mass-energy, then gravity would have won out and the universe would have collapsed after a few million years. On the other hand, if the mass-energy had been less or the cosmological constant— the value for the energy of empty space—even a tiny bit greater, the universe would have been essentially empty after a few million years. In neither case, would life have been able to develop in the universe. Noted physicist, Stephen Hawking, calculated that the odds that these numbers would come about by chance as one in 10^{18}. He concluded that chance alone could not explain the fine-tuning of our universe—there had to be some other explanation.[10]

There are four fundamental forces responsible for shaping the universe. These are gravity, electromagnetic, weak, and strong nuclear forces. The distances covered and strengths of the four forces differ greatly. The strong nuclear force is extremely strong since it must overcome the mutual repulsion of protons in the nucleus of atoms; however, it only acts over an extremely short distance. The nuclear weak force is intermediate in strength and acts over a short distance. The electromagnetic force is also intermediate in strength and acts over a short distance (for example, the electromagnetic attraction of electrons for protons), and over great distances in the form of radiation, such as light. Gravity is an extremely weak force but it can act over astronomical distances.

The exquisite fine-tuning of these four forces is required for a long-lived universe and for the conditions necessary for the emergence of life on Earth. For example, the strong nuclear force, which binds protons and neutrons in the nucleus of atoms, is just the right strength to make stable atomic nuclei. The relative strengths of the strong nuclear force and gravity had to be very finely tuned in order for stars to form and be long-lived. Increasing the strength of the strong nuclear force by even

2 percent would mean that protons would not form from the neutrons formed in the Big Bang and atoms would not exist.

All the various particle masses had to be fine-tuned in order for life to exist in our universe. For example, the proton-neutron masses differ by about 1 part in a thousand and equals twice the electron's mass. If this difference were even slightly different then all neutrons would have quickly decayed into protons or else all protons would have irreversibly decayed into neutrons. In either case, stable hydrogen-burning stars could not exist. In addition, the fusion of hydrogen nuclei to create helium in the center of stars releases just the right amount of energy to power stars for a very long time, thus maintaining a more moderate and constant temperature on planets such as Earth. This enabled higher life forms to evolve on Earth, a process that took billions of years.

If the nuclear weak force had been slightly stronger, then the Big Bang would have burned all hydrogen to helium. Neither water nor long-lived stars would exist. The weak force is also perfectly tuned to allow for the radioactive decay of heavy elements, which heat Earth's molten core, resulting in volcanism and a magnetic field that shields life on Earth from deadly cosmic radiation.

The electromagnetic force is responsible for light, heat, electricity, and for the chemistry that powers life. Even a slight variation in the strength of this force would have a profound effect upon stellar development and the chemistry of life. For example, if the electromagnetic force were slightly different stars might all be red or blue and either too cold or hot to create conditions on planets for the emergence of life. Such stars would also be unable to explode as supernovae and thus create the heavier elements required for life.

The chemistry of life also depends on the fine-tuning of this force and is responsible for the unique properties of water, such as its high freezing and boiling points, solvating power, and expansion upon freezing. These properties have been essential for the evolution of life on Earth. Any slight change in the nature of the electromagnetic forces at work holding molecules together would drastically change the properties of other important molecules needed to support life including proteins and DNA.

If the force of gravity differed even slightly from its actual value, models predict that stars, galaxies, planets, and hence living organisms could not have developed from the dust left by the primordial Big Bang. Even the formation of matter following the Big Bang depended on a tiny excess of matter particles over antimatter particles—something that would not be predicted by current theories. The universe as we know it could not exist without what turned out to be one part in a billion extra particles of matter compared to antimatter particles following the Big Bang.

Another example of fine-tuning is the value for the Higgs field, which is 246 GeV.[11] If this field that confers mass to all the elementary particles was different by even a small amount then the mass of all particles in our universe would be different resulting in the familiar features of physics, chemistry, and biology being different. No doubt, such a universe having different physical properties, constants, and laws would make life as we know it impossible.

The simple fact is that if the Big Bang were rerun with the slightest change in the number, position, or mass of a few particles; a field's strength; a constant's value; or for that matter almost any fiddling at all, then the cosmos that unfolded would not include stars, galaxies, the planet Earth, the human species, you, and I.

Considering the unlikeliness that everything came together spontaneously and with utter perfection in our universe out of the random explosive expansion of the Big Bang, materialists have had to come up with a plausible explanation for fine-tuning. They postulated "multiverse theory" or the theory that there were multiple universes created in the Big Bang. Our particular universe has the perfect balance of factors because an almost infinite number of universes were formed after various regions condensed or emerged from the "inflation" stage of the Big Bang. Fine-tuning forces us to believe in either multiple universes or a Creator—and for materialism the choice is multiple universes.[12]

Inflation and the Multiverse

Inflation theory was first proposed by Alan Guth in 1980. He hypothesized that the universe first underwent a period of extremely rapid expansion (inflation). Following this very short period of hyper-expansion, our universe, and other universes could condense following inflation with slightly different values for the cosmological constant, total mass-energy, physical forces, and constants that would govern it. This would be like the production of a long sheet of bubble wrap where all these universes were separate and did not interact with one another. Thus, while the probability might be extremely small that conditions conducive for life would exist in one of these universes, the almost infinite number of possible universes insures that there must be a finite number of life-supporting universes. As a corollary to the multiverse they argue that we could only exist in a universe that was fine-tuned for our existence—the so-called "anthropic principle."

There is evidence that supports inflation theory, but little if any that supports a multiverse. All the universes are supposedly separate from each other and therefore do not interact with one another. Therefore, it is unlikely that there can be a way to test the theory. A skeptic of materialism might suggest that a multiverse and the anthropic principle are necessary to explain fine-tuning, but conveniently, the hypothesis of a multiverse is untestable.

The New Model

We have seen how many of the postulates of classical materialism have been shown to be false by the discoveries of modern science. Therefore, we might ask whether there is still scope for scientists today to be wedded to the materialist worldview. The answer is of course yes. Many scientists today believe strongly in materialism. However, it is a new version that is consistent with the evidence from quantum physics, relativity theory, cosmogenesis, genetics, and neuroscience. The old ideas that matter is the "ground substance" of reality, locality of matter-energy interactions, determinism, and absence of an effect of human observation on physical reality have had to be thrown out or modified. However, there is still no need to consider any "supernatural" forces, God, or survival of consciousness following death.

Of course, the materialist worldview comes with multiple nuances and no single description could hope to satisfy the view of all of today's materialists. Nonetheless, a few features that best describe the new materialism are listed below.

- The universe emerged spontaneously from nothingness approximately 13.8 billion years ago in what has been termed the Big Bang. A multiverse was formed and our particular universe was one that had all the precise conditions required for life and the evolution of sentient beings.

- Both space and time (spacetime) along with matter and energy began with the Big Bang, and have evolved according to the laws of physics into the cosmos we observe today.
- The initial state of the universe was one of extreme low entropy (great order), and entropy continues to increase driving both the arrow of time and the formation of stars, galaxies, and living organisms. As the entropy of the universe increases, free energy decreases. Eventually the universe will reach an equilibrium state of maximum entropy and zero free energy resulting in the "thermal death of the universe."
- Nonlocality is a fact. For example, when we observe that one electron of a pair of entangled electrons is spinning clockwise (CW) the wave function for the pair collapses and we know that the other electron is spinning counterclockwise (CCW), even if it is in another galaxy.
- The wave function, not matter, is fundamental reality. The physical reality we experience emerges from a subtler or more fundamental domain—that of the wave function. In other words, the basic "stuff" of reality is not composed of atoms and the subatomic particles that make them up.
- Strict causal determinism (the law of cause and effect) does not exist because probabilities and not certainties apply to how physical reality emerges from the quantum realm. However, it does apply to events that occur in the realm of physical reality.
- It is possible that a wave function describes the entire universe, and therefore it is possible that everything in the universe is entangled. When we observe one electron from a pair, our personal wave function (or that of our instrument) becomes entangled with the wave function of the electron. There is a resulting collapse of the wave function for the electrons or a decoherence that breaks the superposition. It is possible that this causes the universe to split (branch), and in one universe, we observe the electron spinning one direction and in the other, we observe it spinning in the other direction.
- We may have an observation problem in that the outcome of a quantum event requires an act of observation, but this does not imply that consciousness or any other nonphysical thing is required. The collapse (or branching) of the wave function can occur by simply placing our instrument into such position that

it can be entangled with the thing we want to measure, or that collapse then occurs randomly similar to radioactive decay.
- Relativity theory accounts for the fact that motion and gravity affect time. The speed of light is a constant but the four-dimensional fabric of space and time (spacetime) can be compressed or expanded depending on the relative velocity or mass of an object. This warping of spacetime by mass gives rise to the force of gravity.
- Life as we know it was formed by random chemical reactions that first produced complex molecules that then interacted to form simple single-celled organisms with the capabilities of reproduction and metabolism. More complex living organisms evolved from these primitive life forms by a process first described by Darwin in which changes that confer a survival advantage are passed on to future generations.
- Evolution is responsible for the complex physical structures, bodies, and nervous systems seen in advanced animals.

3

THE WONDERFUL WORLD OF THE QUANTUM

Some History and Some Experiments

What is Quantum Mechanics?

QUANTUM MECHANICS IS A theory that describes the microscopic world. As we have seen, the word quantum comes from the fact that certain values for the particles (mostly subatomic particles, atoms, and molecules) and the motivating forces or energies involved with them are restricted to discrete values (quantized). In addition, these objects have characteristics of both particles and waves (wave-particle duality), and there are limits to how accurately the value of a physical quantity can be predicted prior to its measurement, given a complete set of initial conditions (the uncertainty principle).

Quantum mechanics has been fantastically successful in describing how atoms and particles interact through the forces of nature, and predictions it has made have been found to be enormously accurate. Probably no other theory has been more responsible for the advancement of modern technology. Such things as transistors, microchips, lasers, and MRI machines depend on quantum mechanics to work.

Quantum mechanics gradually arose from theories to explain observations that could not be reconciled with classical physics. Examples were Max Planck's solution in 1900 to the blackbody radiation problem; Albert Einstein's 1905 paper that explained the photoelectric effect; and from studies of the emission spectrum of atomic hydrogen.

Aside from Planck and Einstein such men as Bohr, Heisenberg, Schrödinger, Born, Dirac, and others had by the mid-1920s developed mathematical formalisms to describe the behavior of quanta. The most important of these, the wave function, provides information about the probability amplitude of energy, momentum, and other physical properties of quanta.

Quantum mechanics (or quantum physics if you prefer) is often described as "mysterious," "difficult," or "incomprehensible." There is an element of truth to this description, but to most physicists it does not matter whether they understand what is going on since the theory works. They find it is easier to use quantum mechanics to solve problems than to worry about what is really going on and what these mysterious experimental observations imply about reality.

One of the greatest mysteries of quantum mechanics is that a quantum system behaves differently when we *observe* it than how it behaves when we are *not observing* it. One possibility is that quantum mechanics is an incomplete theory because it does not explain what an "observation" or "measurement" is and how it changes the system. Perhaps someday a scientist will come up with a mathematical model of the theory that incorporates observation or consciousness in it. The other alternative is that quantum mechanics is telling us that an objective reality that is independent of us *does not exist*. Somehow, observation is fundamental to how reality manifests.

You might assume that physicists would be very concerned about the question of what it all means and spend a lot of time and money in an endeavor to find out. But this is not the case. Most are content simply to use this tool to solve problems and have little interest in understanding how and why quantum mechanics works. The fallback position of most physicists today is to go with the materialist model of the universe and not worry about the "observation problem."

The Key Experiment

If there is a single experiment that illustrates the weird, non-classical nature of quantum theory, it is the double-slit experiment. To quote physicist Richard Feynman, "It has in it the heart of quantum mechanics. In reality, it contains the *only* mystery."[13] Embodied within this one experiment are all of the essential ingredients that reveal the nonlocality of space and time and the effect of observation that makes quantum mechanics so hard to correlate with our everyday experience of reality.

In this experiment, light (photons) or electrons are passed through two slits that are very close together. Both electrons and photons have wave and particle characteristics depending on how they are observed. If we take light for our example, then the light waves passing through the slits interact to form an interference pattern (alternating dark and light lines) as the waves either cancel or reinforce one another. This is exactly what is expected if light behaves like a wave.

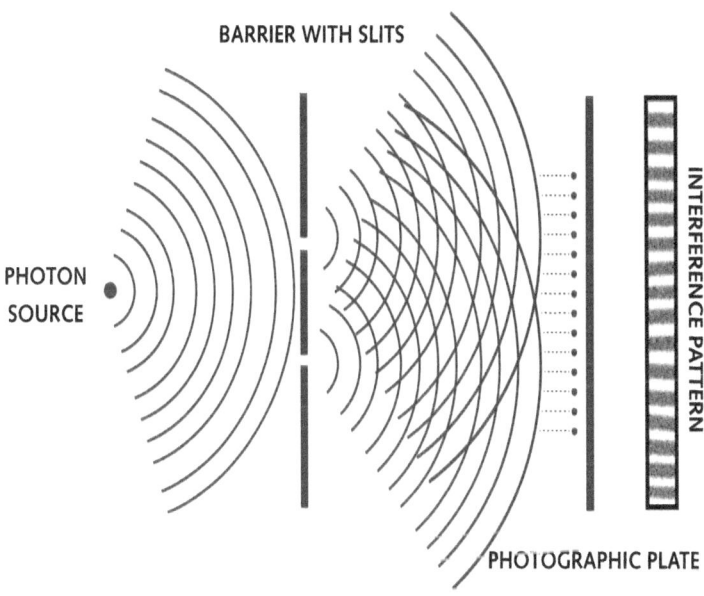

Double-slit experiment. Waves from the photon (light) source pass through two closely spaced slits causing the wave fronts to interfere with each other producing alternating dark and light lines.

However, light can also act like a flow of particles. If the intensity of the light beam is lowered sufficiently then only one photon at a time will pass through a slit of the apparatus. In this case, the interference pattern might be expected to disappear since it is created by the interaction of wave fronts from the two slits, and the individual photons arriving at the slits at different times should not be able to interact. However, the pattern persists even though the individual photons cannot split, each half going through a different slit and then recombining before they hit the detector. The interference pattern may be observed by exposing a photographic plate for many minutes or summing the response from a photoelectric detector. The existence of the interference pattern under these conditions suggests that the so-called "well-defined" particles are behaving like a wave and individual particles are interacting with particles that passed through the slit at different times. That is, they exhibit both spatial and temporal nonlocality.

The double-slit experiment can be modified in various ways that demonstrate different aspects of quantum behavior. For example, electrons, which are considered particle-like in contrast to light, produce an interference pattern just like light. Recently massive molecules with up to 2000 atoms have been shown to produce an interference pattern demonstrating that even these massive particles have wave properties.[14]

Another variation is to set up a detector at one of the slits in order to determine what slit an electron went through. As expected, the detector indicates that only whole electrons go through a slit since like photons they cannot divide into two halves and then recombine before hitting the detector. At the same time, the interference pattern disappears and only a normal diffusion pattern that is typical of a single slit is observed. Somehow, the electron knows that it is being tracked and goes to a different location on the detector.

One of the slits can be monitored making it effectively unavailable to the electron while it is in flight. Again, in this case, only a simple diffraction pattern is observed. Theoretically, if the electron is forced to pass through one slit, it is forbidden from acting like a wave and reverts to its particle nature. Interestingly, the same thing happens when the slit is monitored *after* the electron passes through one of the slits but before it hits the detector. In this so-called "delayed choice" experiment, the disappearance of the interference pattern (the effect) precedes the cause (monitoring of the slit). Hence, the electrons not only behave as though they know they are being watched, but can *change their behavior retroactively*.

THE FALLACY OF MATERIALISM 33

In another version of the experiment, the only change in the experimental design is information that is conveyed to the observer's mind. Here the detector is monitored by a computer that randomizes the results making them incomprehensible to the scientist. In this case, the electron behaves like a wave producing an interference pattern. However, when the randomizer is turned off and the observer gets information about what slit the electron passed through the computer indicates that it has reverted to its particle nature. And this change can occur instantaneously or even days later when the data is analyzed by the computer. Whether the electron behaves like a particle or a wave depends solely on what the scientist knows about it.[15] In other words, it is not the measurement that causes wave function collapse but *information*.

These are examples of temporal nonlocality for quantum phenomena—such phenomena are not bound by linear time. The question is how does the electron that goes through the open slit "know" that the other slit will be closed or monitored after it passes through and that it must go to a different location on the detector? When the other slit is not fully available to it, the electron would appear to be prevented from acting like a wave. Apparently, the set-up of the entire system or apparatus determines the behavior of the electron. Such behavior is clearly connected in some way to the apparatus that is used to detect it and to the intention or mind of the scientist. As mentioned earlier, physicists call this connection entanglement. Such interconnectedness exists anytime two entities share the same wave function. When two or more entities share the same wave function then they can no longer be considered separate.

In other words, all these experiments can be understood in terms of quantum wave mechanics. If the electrons or photons are described by a wave function, then before they are disturbed by some form of observation or measuring device, it is natural for them to act like waves and produce an interference pattern. However, as soon as the wave function of the quantum is confined in any way by a measuring device or human observer with their own wave function there is entanglement of the two wave functions that causes a collapse of the combined wave function to give a particle with specific properties.

The results of the double-slit experiment clearly demonstrate that there is *no such thing as well-defined particles*. One must conclude that all quantum particles have some "fuzziness" as defined by the wave function. The wave function cannot be some abstract aspect of reality since it causes real phenomena such as interference patterns. Instead, we are forced to conclude that it represents reality at a subtler or more fundamental level than that of photons, electrons, atoms, molecules, and physical matter in general.

Experiments have consistently demonstrated that the behavior of quanta is not just determined by the conditions of the test. Their wave-like or particle-like behavior depends on the totality of the experimental apparatus and the intention of the experimenter. The best one can say is that a quantum behaves like a wave before it is observed and like a particle after it is observed. This highlights an important aspect of quantum mechanics—the *observer and the observed system cannot be separated*. The observer or his instruments are part of the system and influence the outcome of the observation. In other words, the act of observing alters or influences the system and this alteration can occur retroactively.

Testing Bell's Theorem

Quantum mechanics, especially the version that was formulated by Neils Bohr and his collaborators in Denmark, required that quantum particles be *nonlocal* in nature.[16] Theoretically they could influence each other instantaneously even at great distances—i.e. be entangled. Moreover, it said that we can never know with certainty both the position and momentum of a quantum particle; at best we can only assign a probability that it will be found at a particular location.

The idea that quantum theory predicted faster-than-light communication between quanta really bugged Einstein. He called it, "spooky action at a distance." He felt that any theory that violated the speed-of-light restriction of special relativity and could not provide predictable quantities to elements of reality such as both position and momentum must be an incomplete theory. His main objection to quantum theory was published in a 1935 paper along with collaborators, Podolsky and Rosen, and is known as the "EPR paradox." EPR argued in favor of the principle of *locality*, which says that particles must exist as specific points in spacetime—where they can only interact with their neighbors and cannot be spread throughout the universe. In addition, they argued that quantum mechanics could not be a complete theory if it failed to provide an explanation for all the elements of physical reality.

The real issue here was separability. The material realism of Einstein considered all objects as separate entities, while the new physics of Bohr specified that quantum particles were nonlocal and could be connected at a level more subtle than spacetime. Einstein maintained that, for example, if one electron in a pair was found to be spinning clockwise when observed, then its partner spinning counterclockwise must have started that way. It would be like sending one glove in a pair to a colleague in Boston. When he opened the box and found a right-handed glove, he would instantly know that the other glove you had was left-handed. There is actually no mysterious entanglement going on, just a theory that is incomplete because it cannot predict the "handedness" of material objects.

The debate on whether the nonlocality predicted by quantum mechanics was correct might still be going on if it were not for the ideas of physicist, John Bell, who in 1964 discovered a way to prove conclusively which version of reality (locality or nonlocality) was correct. It is known as Bell's theorem of inequalities. The theorem concludes that the best a local theory could do is produce a 50 percent correlation between certain spin states. If these could be measured and shown to correlate more than 50 percent of the time then nonlocality (entanglement) had to occur. However, it took another 20 years before the experiments testing Bell's theorem could be conducted.

Probably the first definitive experiment testing Bell's theorem was conducted by Alain Aspect and his colleagues at the University of Paris-Sud in 1982. Aspect set up an apparatus that emitted paired photons. That is, two photons emitted simultaneously from the same source having opposite polarizations. According to quantum mechanics they are born from the same source with the same wave function and are never truly separate—but are entangled, two parts of a whole. If the polarization of one is observed to be vertical, quantum theory requires that the polarization of the other particle must be horizontal. The polarization of the two particles will always be observed to be opposite.

By changing the angle of the polarization detectors, Aspect was able to demonstrate a greater than 50 percent correlation as predicted by Bell if there was a nonlocal connection between the photons. This observation alone did not prove nonlocality since the photons might communicate via local fields or signals at the speed of light. So, Aspect added an electronic polarization switch that had the effect of changing the polarization setting of one of the detectors every ten-billionth of a second—shorter than the time for light to travel between the two detectors. It turned out that the change of the polarization setting in one detector influenced the outcome of the measurement in the second detector showing that the two photons were indeed connected through space just as predicted by quantum theory. There was just no time for the information about the change in one detector to be transmitted to the other paired photon through local signals.

Practical Uses of Entanglement

Numerous and even more definitive experiments have been conducted since the 1980s confirming entanglement as predicted by Bell's theorem and demonstrating beyond any doubt that entanglement is a fact of nature. Today researchers are using quantum entanglement to run quantum computers—quite possibly a breakthrough in computer technology. Quantum computers use quantum bits (qubits) instead of bits. A bit is either a one or zero, but a qubit is a system that has two possible measurement outcomes—for example, spins CW or CCW. However, the qubit is a superposition of these spin states and can become either upon being measured or manipulated by the computer. Unlike a normal bit, which is fixed at a value of one or zero, the qubit can assume either of the two possible states. Because qubits can exhibit quantum entanglement, a set of qubits can express higher correlation than is possible in a classical system. As a result, a quantum computer can theoretically have orders of magnitude greater computing power than even the most powerful classical computers in use today. Although the development of such "supercomputers" is still in its infancy, researchers at IBM have successfully built a prototype processor having fifty qubits. While fifty bits for a normal computer is equivalent to seven bytes and could not even code for the word "computer," in quantum physics one requires 2^n bits to describe the system completely. For just fifty qubits, this could theoretically be equivalent to 10^{12} bits (one hundred terabytes) of data, which is equivalent to ten times the print collection of the U.S. Library of Congress.

A quantum computer with its entangled qubits, which can be one, zero, or either is theoretically able to make many calculations at the same time unlike a classical computer that can only make a series of computations—albeit very rapidly. It could be compared to putting a single rat into a complicated maze with hundreds of dead ends and only one way out vs. putting a hundred rats into the maze at the same time. Surely, in the latter scenario one of the rats, by chance will find the elusive escape route in a short time. In the future, quantum computers will likely be developed that will greatly enhance robotics, artificial intelligence, and solve many complex problems such as weather and financial forecasting that are not possible using today's classical computers.

Quantum communication is another technology made possible by entanglement, and it has already been demonstrated by Chinese researchers using a space laser.[17] We might assume that such communication could be instantaneous because information about the state of one entangled particle can be instantaneously communicated to its twin. However, it is not possible to start with a quantum superposition and force it into a single state. As a result, no useful information can be communicated faster than the speed of light. However, entangled photons can encode information in such a way that it would be impossible to break an encryption code. Normally if you want to send a coded message between two people, you must give them both a secure key that allows them to translate the message. At the same time, you must protect that key from any nosy third parties who are trying to spy on the conversation. A complex quantum key, shared via entangled particles, would do the trick, because if a spy tried to steal the code-breaking information, this would disrupt the entanglement making it useless for the intruder. The beauty of quantum communication is that the integrity of the data sent is protected by the laws of physics, and thus there can be no higher level of security.

Quantum teleportation is similar to quantum communication in that quantum information, such as the exact state of a photon, electron, ion, or atom, is transmitted from one location to another. It is another well-established example of quantum entanglement. It differs in that quantum teleportation provides a mechanism for moving qubits from one location to another without physically moving the underlying qubit particles. Quantum teleportation can take place when there is previously established quantum entanglement of particles at the sending and receiving locations, and the information about the particles is sent by way of a "quantum channel" to the receiving station from the sending station. In the process of transfer, the information carried by the particle at the sending station is destroyed. While the word "teleportation" conjures up images from Star Trek, it cannot be used to transport material objects—only information. No doubt, it would take enormous technological advances before this quantum information could be used to assemble even a simple object.

4

INTERPRETATIONS OF QUANTUM MECHANICS

QUANTA DO NOT FOLLOW the rules of classical mechanics and the behavior they exhibit suggests that we might need a new model of reality. This need has spawned a new field of study that may better fall under the umbrella of metaphysics than physics termed the interpretations of quantum mechanics.[18]

The philosophical implications of quantum mechanics have never been a popular topic of study among physicists. Speculating about the nature of quantum reality could essentially end one's career in physics. However, a few brave souls have persisted in this endeavor and we have today several different theories on what quantum mechanics implies about reality.

The Copenhagen Interpretation

As the name implies, this is the interpretation of quantum mechanics first proposed and developed by Neils Bohr and his associates in Denmark. It is the oldest of the proposed interpretations of quantum mechanics, and it is the "textbook" version that is commonly taught to physics students.

According to the Copenhagen interpretation (CI), quanta do not have definite properties prior to being measured. Any of their properties such as position or spin are best described by the wave function. This function evolves as described by the Schrödinger equation to produce certain allowed or possible outcomes that will occur upon observation. The probability of getting any one particular result can be calculated by squaring the amplitude of the wave function corresponding to that outcome (Born Rule). Upon observation or measurement, the wave function *collapses* resulting in a specific result. As mentioned previously, this feature is known as wave function collapse. Before measurement, the wave function for position would be nonlocal or spread throughout the entire universe, but after measurement, it is reduced to a specific point in spacetime.

This is what is observed, for example, in the double-slit experiment. Apparently, nature works this way, but the implications of wave function collapse by measurement or observation are enormous. It means that a macroscopic entity such as a camera or electron detector causes the wave function of a microscopic entity to emerge into physical reality in a specific state. In the absence of such an interaction, the system will merely remain a superposition of all possible outcomes as defined by its wave function. We could explain the interaction between the microscopic and macroscopic entities by saying that a macroscopic entity (e.g. a detector) also has a wave function, and it can become entangled with the wave function of the microscopic entity (quantum). When entanglement occurs, there are not two separate wave functions, but now a single wave function describing the whole "system" of particle and detector.[19] Like any other wave function, this one would happily go on being a superposition until it is collapsed by an observer that is somehow outside the system.

Now consider that the human being that set up the detector would have their own wave function that could be entangled with the quantum and the detector. The wave function for the macroscopic entity would have to include the one for a human being that set up the system. Collapse of this composite wave function must also include an observer. However, this observer must be *outside* the combined superposition of the quantum, the device, and the human. The only thing that would satisfy this situation is a nonphysical entity that we might label mind or consciousness. Hence, the enormous problem of explaining this "observation problem" is that any form of measurement we can think of involves human consciousness or intent in some way.

Ultimately, a nonmaterial aspect of a human being would have to be involved in wave function collapse. Otherwise, there would be nothing that could be considered outside the system. We are therefore left with only one possible explanation—if the Copenhagen interpretation of quantum mechanics is correct: *collapse of the wave function must be caused by a nonphysical entity*. And this entity operates through human beings. Thus, wave function collapse is caused by mind/consciousness and these must be nonphysical.

Probably the majority of physicists today reject the idea that consciousness should have any role in the manifestation of reality. They would prefer not to deal with metaphysical questions and stick with scientific materialism—if possible. Such physicists are more inclined to believe in some of the other interpretations of quantum mechanics that sidestep the requirement that human observation is required before physical reality manifests from the subtle realm described by the wave function. However, as we will see later in this book, it makes a lot more sense, considering the totality of the evidence, to include consciousness in the equation.

The Copenhagen interpretation of quantum mechanics can be summarized as follows.

- A quantum system exists as a superposition of possibilities as defined by a wave function (Ψ).
- The system evolves deterministically according to the Schrödinger equation.
- There are certain properties of the system, such as spin that can be measured but the probability of getting any particular result can be calculated from the wave function.
- The act of observation or measurement collapses the wave function giving one of the possible results—the other possibilities disappear.
- A particular experiment can demonstrate particle or wave properties for quanta, but not both at the same time (Bohr's complementarity principle). All matter and energy possess both particle and wave behaviors, although for large objects the wave nature is insignificant and can be ignored.
- Separability does not exist in the wave function. It is nonlocal and exists not in spacetime but in abstract "configuration space" that can be considered a subtler or more fundamental level of reality.

Below are some of the implications of this interpretation.

- Beneath the classical universe we experience lies a subtle realm of wholeness from which reality manifests upon our observation. Separate parts, just as separate particles, emerge only upon an act of observation.

- Our mind is an observer-participant in the creation of the reality we experience. The change from potentiality to actuality comes through our mental intention and corresponds to a change in our knowledge of the world (information).
- Reality is more mind-like than matter-like. It appears that mind is necessary for matter to emerge into the physical realm suggesting that matter is actually an epiphenomenon of mind.

It is easy to understand why physicists that favor the materialist worldview would be opposed to the CI since it indicates that something nonphysical (i.e. mind) is needed to initiate collapse of the wave function. These implications of the CI are an outgrowth of the need for observation to cause wave function collapse. In the process, the other possible states described by the original superposition disappear. But what if we can eliminate the collapse? Then perhaps the observation problem can be solved? There might be a cost to pay for this idea, but it might remove consciousness from the equation.

The Many-worlds Interpretation of Quantum Mechanics

The many-worlds interpretation (MWI) of quantum mechanics was first proposed by Hugh Everett III in the 1950s. It postulates that every time we make an observation the world branches, and there is another copy made—one in which, for example, the electron is spinning clockwise and another where it is spinning counterclockwise. Therefore each of us exists in countless worlds, all real, but slightly different. It retains some determinism because the world we observe is the branch of the world that evolved according to the "rules" of cause and effect—not some random process.

It postulates that a wave function describes the entire universe. This wave function is the basis of physical reality. Physical reality evolves from it and everything in the universe can be considered entangled. When we observe an electron, which could have spin CW or CCW, our personal wave function (or that of our instrument) becomes entangled with the wave function of the electron. There is no actual collapse of the wave function for the electron, but a *splitting of the superposition along with a splitting of the universe.* In one universe, we observe the electron spinning CW and in the other, we observe it spinning CCW. Both universes are equally real and evolve according to the Schrödinger wave equation. The two universes cannot interact with one another and therefore it is impossible to prove that they exist. All universes just go on evolving and splitting indefinitely forming an almost infinite number of universes—all with slightly different outcomes.

 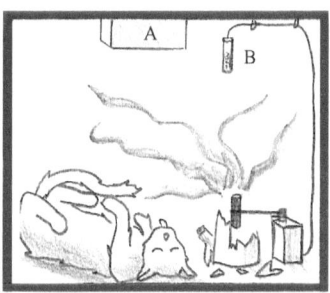

To illustrate this we can use the example of the well-known thought experiment, which has been dubbed "Schrödinger's Cat." In this scenario the behavior of a quantum particle is linked to the life or death of a macroscopic entity—namely a cat. The cat is put in a box along with a trace amount of radioactive substance in A that might omit a beta particle every hour or so, along with a Geiger counter at B that is linked to a hammer that will break a vial of poison. There is no way of knowing when the substance will undergo radioactive decay and hence whether the cat is alive or dead without looking in the box.

Quantum indeterminacy requires that there exist a superposition of two states—one in which the cat is alive and the other in which the cat is dead. In a sense we can say that in one superposition it is half-alive and in the other half-dead. The superposition is broken only when we look inside the box. According to the Copenhagen interpretation, a system can exist in two or more states simultaneously until the superposition collapses upon observation and only one of the possible states is observed.

With the MWI, the other state of the cat that is not observed does not disappear but becomes a new universe. Both universes are equally real and continue to evolve in a normal manner. We might imagine that in one universe we go on living with our cat to a ripe old age, while in the other universe we die young because a mouse gets into our house carrying a deadly disease.

In MWI causal determinism still occurs. The outcome of any event is completely determined by prior states. The only problem is that there is still some uncertainty in our measurements and thus a small degree of randomness. Furthermore, we cannot predict in advance which universe we will end up in for any given observation. Hence, our ability to predict an event such as the spin of an electron—even knowing all prior information about the system—is still only 50 percent.

In the MWI, observation is still required before a quantum system emerges from the realm of possibilities described by the wave function, but this does not imply that a nonphysical thing like mind is required. There is no wave function collapse but a branching. This can occur anytime a measuring system interacts with a superposition and then this combo system interacts with the environment in a process call decoherence. There is in reality a single wave function for the universe and therefore everything in the universe is entangled, but for practical purposes, we can assign subset wave functions to a quantum and to a measuring device. Although the quantum's wave function is simple and might describe only two states like spin CW and spin CCW, the measuring device has a complicated wave function and is entangled with the environment and even the experimenter. When the quantum and the detector interact during a measurement, decoherence inevitably takes place.[20] The result is a branching of the universe. After the branching, there are two copies of the original observer and the universe.

Since such branching is occurring all the time, we might ask how many versions of the universe there are along with versions of ourselves. There is no definite way to answer this question, but we would surely exist in countless worlds, each with a slightly different version of ourselves.

Some physicists like the MWI of quantum mechanics because it maintains determinism and is a simple and a straightforward consequence of applying the Schrödinger equation. For example, Caltech physicist Sean Carroll, an outspoken proponent of MWI, calls it "lean and mean."[21] There is no need for one of the branches to disappear, but each branch of the many universes still requires a measurement before splitting can occur. Since measurement implies an observer, it is not entirely clear that consciousness can be excluded from the model.

It is no surprise that physicists that favor a materialist ontology might also favor this interpretation of quantum mechanics. It describes reality in an objective manner, and on the surface, it does not require mind or consciousness to cause wave function collapse. However, it does have some unsettling consequences about the nature of reality. These are outlined below.

- This universe is constantly splitting into a stupendous number of branches (perhaps as many as 10^{100})—resulting from measurement-like interactions between the myriad components of the universe. Moreover, every quantum transition that takes place in every corner of the universe is splitting our local world on Earth into countless copies of itself.
- Although MWI predicts an awful lot of universes, these cannot interact with one another, and therefore there is no way to know if they actually exist. Hence, it is untestable and unfalsifiable—something science does not like.
- MWI denies a singular self. Our notion of self is also split, and we exist in multiple worlds. Our self can be considered a tree with countless branches. Each version is slightly different with different life experiences and outcomes. The further we go from this moment in time the greater the differences in the life histories of our various selves. Our perception of only existing in one universe is illusory. The reality is that we exist in countless worlds.
- The wave function is real—not just a theoretical thing in abstract mathematical space. We might think of atoms as real, but true physical reality is the wave function.[22] In addition there is a wave function for the universe and it is real.[23] Therefore, everything in the universe is entangled.
- MWI appears to violate Occam's razor, which states "entities are not to be multiplied without necessity." Adherents of the MWI argue that on the contrary it is a simplified interpretation of quantum mechanics because it eliminates the disappearance of the unobserved possibilities described by the wave function.
- MWI presupposes a multiplicity interpretation for quantum superpositions. The basic assumption being made is that quantum superposition by default equals multiplicity. However, when you use ordinary probabilities, you are not obligated to believe that every outcome exists somewhere. For example, an electron's wave function really may be describing a single object in a single state, rather than a multiplicity of them.
- Since each branch of the many-worlds requires a measurement

at the point of bifurcation or splitting, measurement implies an observer. Indeed, human observation does cause splitting. The "observation problem" is not really solved by saying that measurements can occur spontaneously. It also should go without saying that spontaneous splitting has never been observed.

MWI tries to eliminate the need for a human element in bringing physical objects into reality. However, because it is unfalsifiable with no scientific evidence supporting it, and because it denies a singular self, MWI ontology falls under the category of a metaphysical oddity rather than science. Finally, by introducing a universal wave function as the foundation for reality, it may contribute to quantum cosmology, but at the same time may have the unintended consequence of arguing in favor of a cosmology of *oneness*. This is something that is discussed further in Chapter 6.

Many physicists have questioned the veracity of this theory of reality including John A. Wheeler, who was Hugh Everett's PhD advisor at Princeton.[24] Physicist Paul Davies objects to the MWI because of its unnecessary complexity, and because it is hard to see how law and rationality can emerge from total randomness.[25]

Bohmian Mechanics

Another approach to the measurement problem is exemplified by what can be termed "Bohmian mechanics" because it was formulated by physicist David Bohm. Suppose that the wave function does not provide the full picture, but that there are hidden variables. This sounds like Einstein's objection to quantum mechanics, but Einstein hated this interpretation because it depends explicitly on nonlocality. The variable that is hidden in Bohmian mechanics is a "guiding equation," which depends on the configuration of the system given by its wave function. Like MWI, the theory is deterministic, nonlocal, and reproduces all the experimental observations and predictions of quantum mechanics. It also solves the measurement problem by assigning a wave function to a measuring instrument that evolves similar to the many-worlds interpretation but does not create alternate worlds because the hidden variable describing the apparatus will better correspond to one of the branches than any other branch.

Unlike the CI, Bohmian mechanics does away with the uncertainty or fuzziness describing the position and trajectories of particles. The uncertainty observed in experiments is due to statistical uncertainty regarding the initial conditions of any specific particle and therefore how the additional or "hidden" quantum force acts. In the end, this new picture gives the exact same quantitative predictions as other interpretations.

Bohm's interpretation of quantum mechanics went on to describe reality as having three layers. The crudest layer that contains matter and spacetime he called the explicate order. The subtler level of reality that permeates our four-dimensional world of space and time, he called the implicate order. The hidden realm of the implicate order—identical to that of the wave function—unfolds into the manifest realm of reality and is whole. He argued, "...that ultimately, the entire universe (with all its particles, including the constituents of human beings, their laboratories, observing instruments, etc.) has to be understood as a single undivided whole, in which analysis into separately and independently existent parts has no fundamental status."[26] He also stated, "One is led to a new notion of unbroken wholeness, which denies the classical analyzability of the world into separately, and independently existing parts."[27]

Bohm's concept of the implicate and explicate order is his way of describing the relationship between the material and mental levels of reality. Mind (implicate order) manifests as matter (explicate order), and this is a continuous process of folding and unfolding.

Bohm recognized that the most fundamental level of reality must be consciousness (superimplicate order). For example, he wrote, "Deep down the consciousness of mankind is one. This is a virtual certainty because even in the vacuum, matter is one; and if we don't see this it's because we are blinding ourselves to it."[28]

There is a lot to like about Bohmian mechanics and philosophy even though this interpretation of quantum mechanics does introduce a new variable and describes particles as real when they unfold from the hidden realm of the wave function or implicate order.

Quantum Bayesianism

One of the more interesting interpretations of quantum mechanics is a subjective ontology called quantum Bayesianism (abbreviated QBism). The term Bayesianism comes from Bayes's theorem—a way of describing the probability of an event based on prior knowledge of conditions that might be related to the event. For example, if there is a higher probability that a smoker will develop heart disease than a nonsmoker, then using Bayes's theorem, we conclude that a history of smoking can be used to more accurately assess the probability of someone getting heart disease than can be done without this knowledge.

In this interpretation, the wave function is not considered real as in other interpretations. The wave function is useful for calculating probabilities of experimental outcomes, but the true reality is our experience of the quantum phenomenon. When enough observations are made of a quantum system then a probability distribution will be obtained that matches the one predicted by squaring the wave function. Therefore, in QBism there is no need for wave function collapse because reality is not determined by a singular observer but only emerges from the totality of what multiple observers experience. This can be termed "Participatory Realism," it gets around the problem that a macroscopic object such as the moon does not exist if no one is around to observe it.

There are some things to like about QBism. It defines reality in terms of experience rather than a separately existing outside world. As we will discuss in Chapter 8, we cannot know if physical reality exists using our brain because everything we know about physical reality is based on our experience. Secondly, it is consistent with the idea that a collection of minds (or a collective mind) can bring reality forth from the hidden domain of the wave function.

John Wheeler's Participatory Universe

John Archibald Wheeler (1911-2008) was one of the greatest theoretical physicists of all time. He was a colleague of both Einstein and Bohr and was a professor emeritus at Princeton. His list of accomplishments in physics is impressive, but he was also a scientist-philosopher with an interest in understanding the nature of reality.

Wheeler's argument was that reality is created by observers in the universe. Since observers are an important ingredient in quantum theory, they must play an indispensable creative role in the creation of the universe while at the same time being a product of the very universe they help create. He wrote:

> We could not even imagine a universe that did not somewhere

and for some stretch of time contain observers because the very building materials of the universe are these acts of observer-participancy.[29]

He questioned, "How does something arise from nothing?"[30] Information must be the origin for all things physical, and therefore physical reality can be seen as arising from yes-or-no questions or binary choices obtained from observers or their measuring devices. If the foundation for the physical world is information-theoretic in origin, and information is nonphysical then this universe must be a participatory universe created by something nonphysical, that is, consciousness.[31]

In other words, the physicist observing an electron with her instrument is putting a question to nature and in the process "creating" the electron. Wheeler summed this up as follows:

> No elementary phenomenon is a phenomenon until it is a registered (observed) phenomenon. Useful as it is under everyday circumstances to say that the world exists "out there" independent of us, that view can no longer be upheld.[32]

To Wheeler the single most important and fundamental aspect of quantum mechanics was the requirement of an act of observation, which by necessity implies participation by an observer. He stated this as follows:

> The choice one makes about what he observes makes an irretrievable difference in what he finds. The observer is elevated from "observer" to "participator." [33]

He then asked the question of what this means about our understanding of reality:

> Is not the central role of the observer in quantum mechanics the most important clue we have to answering the question? Except it be that observership brings the universe into being what other way is there to understand the clue?[34]

According to Wheeler quantum theory has opened the door to a new understanding of reality. The observer-participant creates both the physical laws and the appearance of the material world in which the laws apply. Such an observer-participant could be human consciousness, but more logically, the creation of the cosmos would entail limitless cosmic consciousness or cosmic mind. If this were not so, where would one draw the line on observership? Would a mouse fit the bill, a worm, or an amoeba? How about a rock—it might be composed of consciousness in its crudest form? Where in the hierarchy of conscious entities that might populate the universe, would one draw a line and say this type of entity and not that one has the capacity to collapse a wave function? No, it would appear that if observership is indeed required to bring about physical reality then it must be a function of the entire cosmos and not any particular part of it. It is as though the entire universe is the observer and is involved and participating in its own creation.

Wheeler and Time

Wheeler came up with the idea of the delayed choice experiment in which one of the slits in the double-slit experiment is closed after an electron passes through one of the slits but before it hits the detector. Somehow, the electron knows that the slit will be closed and goes to a different location on the detector. Experiments have proven that this occurs and it demonstrates that an effect can precede the cause. Hence, it is no longer tenable to say time flows via a sequence of events from the past to the present, to the future. And if you think that this applies only to quantum phenomena consider the fact that even macroscopic objects must be considered to have wave functions that describe them. The same rule would apply. This is the reason that Wheeler argues that an event in the past such as the Big Bang might not have occurred or been a real phenomenon that shaped our universe if a consciousness had not been an observer of it. In other words, Wheeler is saying that in the absence of such an observer, the universe would exist in potential form until observers came along and made it real—just like the electron when it goes to a different place depending on whether it is observed. This illustrates an important element of Wheeler's ontology—backward causality. This could also be termed "top down" causality. Observation creates reality, which is the same as saying conscious awareness is the cause of reality. Clearly Wheeler is adopting the basic postulate of the spiritual worldview that consciousness, not matter, is ultimate reality—the source from which everything manifests.

Top-down causality can also be understood by saying that time can move backward. Again, consider the double-slit experiment. The experimenter has an idea to measure a photon emitted from an atom that is in an excited state. The photon exists in a potential state until it is created by the experimenter's equipment. This is the backward-in-time leg of the process. Depending on how the equipment is set up, it triggers the expected response on his photomultiplier tube and in his mind in the forward movement of time. The process begins with mind and ends with mind.

In this experiment, the jump back in time is small, but consider the experiment in which a distant quasar appears to be split into two objects by the bending of light from an intervening galaxy between Earth and the quasar. The light that is bent has roughly fifty thousand light years more distance to travel than the light that comes to Earth directly. However, the photon beams from the quasar interfere with each other in exactly the same way as if they were emitted seconds apart in the laboratory. It appears that our observation of the photons causes them to be entangled, despite the fact that they were emitted billions of years ago and arrive fifty thousand years apart.

Other Interpretations

In addition to the more popular interpretations of quantum mechanics mentioned above are several other formulations. In some theories, the Schrödinger equation is modified. In others, wave functions on very rare occasions undergo spontaneous collapse. In a macroscopic object, having many, many atoms this will always occur eliminating the need for a human observer, thereby solving the observation problem.

5

THE MYSTERIOUS DOMAIN OF THE WAVE FUNCTION

So far, there has been a lot of discussion about the quantum wave function without getting into much detail about what it is and what it may indicate about reality. Now it is time to try to get a better handle on what a wave function actually represents.

What is a Wave Function Anyway?

A typical wave may be represented by a sine wave. It has amplitude that oscillates smoothly up and down from an equilibrium point. The amplitude may be considered both positive (above the resting value) and negative (below). For normal physical waves like those for sound, light, or water, the wave propagates energy. The energy corresponds to the amplitude of the wave and to its velocity. For electromagnetic waves such as light in which the waves propagate at the speed of light, the energy is related directly to the frequency of the wave (and thus inversely to the wavelength), giving rise to various types of radiation such as radio waves, microwaves, infrared, visible, and ultraviolet light, X-rays, and gamma rays. While most waves require a medium to travel through—for example, air for sound, water, or a violin string—electromagnetic waves can travel through empty space because they are propagated by electric and magnetic fields that are at right angles to one another. Recall that light sometimes behaves as if it is quantized. These massless particles are called photons, and they are observed for all forms of electromagnetic radiation.

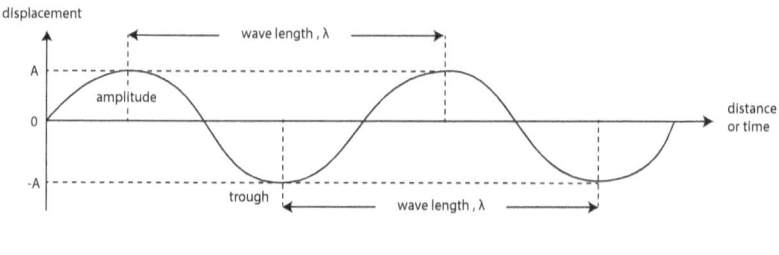

parts of wave

On the other hand, quantum waves are vibrations of possibilities rather than energies or material substances. They are described as having complex-valued probability amplitudes. This is a mouthful and requires some high school mathematics to understand. The fact that the amplitudes of the waves correspond to probabilities is not the hard part. We have already seen that the probabilities for the possible results of measurements made on the system can be derived from the wave function. But, what are complex-valued amplitudes? It turns out that complex in this case does not mean that the mathematics is complicated, just that complex numbers are needed. You might have been introduced to complex numbers in high school algebra, but probably most of us have forgotten about them. A complex number is any number that consists of a combination of a "real" and an "imaginary part." Real numbers are any of the positive or negative numbers or fractions we know and love. However, imaginary numbers contain the imaginary number i that is defined as the square root of -1. Normally, we would think of the square root of a negative number as meaningless, because squaring a number entails multiplying it by itself which must always give a positive number. But there are numerous mathematical equations where imaginary numbers are useful including some equations for Einstein's theory of relativity. If there is any beauty to using imaginary (i.e. complex numbers), it is that when you square them you get a real number. For example $i^2 = -1$.

Wave functions, which are normally symbolized by the Greek letter psi (Ψ), contain imaginary numbers, and when they are squared, they give a probability distribution for where you might find a quantum particle upon observation (Born rule). The real value of the wave function is that it can provide a probability that any of the possibilities described by the function will emerge into physical reality upon observation.

There are real and imaginary parts in the Schrödinger equation as well, and these interact to contribute to the physical results obtained when probabilities are calculated. As a result, there is no getting around the need for a complex wave function. Since imaginary numbers play an important role in both quantum mechanics and relativity theory, they might hold a clue to an understanding of reality.

Wave Function and Wave Equation—What's the Difference?

It is easy to be confused by terms such as wave function and Schrödinger (wave) equation. A wave function is similar to a sine wave, but having amplitudes that can be imaginary numbers. What the wave function is describing is the collection of possible states for a quantum system. Hence, it is a mathematical representation of the quantum superposition. On the other hand, the Schrödinger equation is the basic equation of quantum mechanics and its solution is the wave function. It is used not only to define the wave function, but to determine how the wave function evolves with time. In other words, it also measures the rate of change of the wave function. The rate of change is proportional to the energy and thus higher energy states evolve more quickly than lower energy states.

There is no probabilistic element to the Schrödinger equation. Its description of how the wave function changes is both continuous and causal. It is completely deterministic, but only in describing how the collection of states described by the wave function evolves. Absent an observation, the system will evolve deterministically according to the Schrödinger equation, but it will remain a superposition of many states. It is only after an act of observation that any of the possible states described by the equation will come into three-dimensional reality. Probability or indeterminism comes into play when the system is observed. Prior to our observation, none of the particular states that are possible is elevated to the level of reality.

Even though there is a time factor in the Schrödinger equation, it gives the same result whether time runs forward or backwards. It is only after collapse of the wave function that the reversibility of time is lost. Hence, the domain of the wave function is timeless; the arrow of time begins with an act of observation when the quantum object ceases being spread throughout the universe (nonlocal) and becomes concentrated at a particular point.

Complementarity and the Wave Function

The wave function provides an explanation for things like wave-particle duality. Before we look, things behave like waves. When we look, they behave like particles. Wave-particle duality is an example of complementarity, a principle formulated by Niels Bohr—one of the founding fathers of quantum mechanics. For light, neither description of wave or particle works under all experimental conditions—a complete description requires that we consider both of these mutually exclusive constructs. In addition, our knowledge of the situation is limited because we are unable to measure both constructs simultaneously or describe them precisely (uncertainty principle). In the quantum realm, complementarity is required to describe the entangled and ambivalent qualities of quantum phenomena.

The existence of complementary aspects for describing a physical entity indicates that the true reality of the entity is to be found at a deeper level than that of the two complementary descriptions, and the wave function serves this purpose. The wave function provides the possibilities and observation causes one of the possibilities to manifest in physical reality.

Complementarity also has applications in comprehending and acquiring scientific knowledge in other facets of the physical realm where there is no simple way to describe reality. These include matter/energy, space/time, position/momentum, cause/effect, observer/observed, part/whole, microcosm/macrocosm.

Is the Wave Function Real?

Does it make sense that science might consider the wave function to represent a fundamental reality? This function only speaks of possibilities and says that prior to observation a quantum particle is spread throughout the entire universe. Despite these and other implications outlined below most of the interpretations of quantum mechanics have no problem assigning reality to this hidden domain. This is based on an idea central to science, called "reductionism"— things should be reduced to their simplest and most fundamental nature possible. For example, if we believe that atoms are real and find they are made of subatomic particles such as protons and neutrons, then we must assume that these are real. Then if we find that quarks are the building blocks of these same subatomic particles then we have no choice but to label these as real as well. Since reductionism is a fundamental postulate of materialism it follows that if any of the building blocks of what is considered physical reality emerge from a subtler domain (e.g. the wave function), we must consider that domain to be real.

Implications of Considering the Wave Function as Real

Quantum mechanics has been described by physicists using such adjectives as "weird," "crazy," "bizarre," and "inexplicable." What they are really describing here is the quantum domain—the domain of the wave function. This domain is one of wholeness and such a concept is foreign to Western thinking. Neither science nor Western religions have introduced the concept of wholeness as a feature of reality. As a result, the common-sense experience of separateness is accepted as a fact of nature. The wholeness revealed by the quantum domain is so contrary to our ordinary view of reality, it is no surprise that scientists would use such terms to describe this domain.

However, by all indications the wave function does represents reality at a *more fundamental level* than the one we associate with the objective physical world of atoms, bacteria, planets, and stars—what we normally call physical reality. And it *cannot be a domain of separate parts*, but one in which all possibilities coexist in a state of wholeness—everything being interconnected and interdependent. By necessity, this web of connectivity must permeate the entire universe. The subtle realm of the quantum wave function is thus one of both temporal and spatial nonlocality. Below is a summary of the properties of this hidden realm.

- *The wave function represents a fundamental reality.* It exists at a more basic level than the ordinary reality we associate with our everyday world.

- *It expresses wholeness.* It describes the universe as an indivisible unbroken whole.
- *It is timeless.* The past, present, and future are meaningless when discussing this realm. It is only after the wave function collapses by observation that an arrow of time comes into existence.
- *It is nonlocal.* It penetrates and surrounds ordinary physical reality and is not localized in any part of space but is all-encompassing, everywhere at the same time. It is only after an observation that a "part" of reality represented by the function becomes localized in space.
- *It is a mathematical representation of possibilities.* It determines the probability that any particular quantum possibility will become "real" when observed.
- *Since the wave function describes only potentialities, some outside agent is required to select which of the possible outcomes will manifest in physical reality.* This agent may be labeled mind or consciousness; however, such an agent should not be considered restricted to sentient beings, but could be cosmic in origin.
- *Complex numbers are needed to describe this domain.*
- *The domain of the wave function does not contain energy as such.* Instead, it is the underlying source of all energy. In a sense, it contains the potential for expression of almost infinite energy.
- *The collapse of the wave function does not require energy—just observation.* This is a mechanism by which a nonphysical mind could affect matter (brain)—conservation of energy is not an issue.
- *Theoretically, all matter and energy have associated wave functions.* This includes the brain, the body, and the universe as a whole. The wave function represents the gestalts for these individual entities.
- *Quanta that interact with one another (in either the past or future) must share the same wave function and are entangled.* Since the universe began as a singularity (Big Bang) about 13.8 billion years ago, it follows that everything in the universe is entangled.
- *The wave function that describes the entire universe is fundamental reality.* Ordinary physical reality emerges from this more fundamental aspect of reality. From the spiritual point of view, this wave function is called *cosmic mind.*

6

RELATIVITY THEORY

And What it Implies about Reality

IN 1887, TWO AMERICAN scientists, Albert Michelson and Edward Morley, showed that the speed of light was a constant whether they measured it in the direction the Earth was moving or at right angles to its motion. At the time, all known waves required a medium in which to travel. So most scientists believed that light, which had wave-like properties, required a medium for its propagation through empty space. They assumed that space was filled with an invisible material with no interaction with physical objects, which they called the "luminiferous aether." However, the discovery of Michelson and Morley that the speed of light was not affected by motion suggested that the aether did not exist—a finding that was confirmed in subsequent experiments.[35]

Special Theory of Relativity

Albert Einstein realized that if the speed of light was a constant no matter what point of reference was used, then something else had to change to account for its constancy. He sensed that this *something* must be space itself. He proposed that space could flex and change, become compressed, or expanded according to the relative motion of an object and an observer. The only constant was the speed of light itself or an integrated four-dimensional fabric he called spacetime. These insights led Einstein to propose his special theory of relativity, which states that the universe has four dimensions. There are three of space—width, length, and height—and one of time. Time is not a separate dimension in this scheme, but is fully integrated with the three spatial dimensions. Hence, each of the four dimensions of spacetime has a spatial and temporal component, which is required from the fact that both space and time are relative to the state of motion. Einstein theorized that with motion, space shrinks and time dilates, while for an object at rest, the movement through spacetime is in time alone.

Einstein's equations indicated that the faster an object moves, the slower the passage of time and the more mass it gains. Ultimately, at the speed of light, time stops. However, for matter it would be impossible to attain this speed since it would require all the mass-energy of the universe to accelerate a material object to the speed of light. Subsequent experiments have proven Einstein's theories about space, time, energy, and mass to be correct. For example, the rate of decay of unstable subatomic particles is slowed exactly as Einstein predicted when they are accelerated near the speed of light in a cyclotron. Secondly, such particles gain the exact amount of mass predicted by the theory as they race around in the accelerator.

However, photons, which carry electromagnetic radiation such as visible light, can move at the speed of light since they have no mass. Their internal clocks are stopped and they do not decay like other particles. As an object approaches the speed of light, both the object and space become compressed. For light, this would correspond to compressing space to a point. Light cannot move through spacetime any faster than the speed of light since it is limited by the compressibility of space. Space can be compressed down to a point, but no further. Hence, not only is it impossible for any physical object to reach the speed of light, it is also impossible for light to exceed that speed.

Einstein's special theory of relativity also postulated the equivalence of mass and energy and gave us the famous equation that energy is equal to mass times the speed of light in a vacuum squared ($E=mc^2$). This equation demonstrates that matter is a condensed form of energy, and today physicists express the mass of subatomic particles in terms of energy.

General Theory of Relativity

Einstein's second theory of relativity, termed the general theory of relativity, describes gravity as a geometric property of spacetime. This theory predicted that mass distorts spacetime and that this distortion is what causes gravity. The more massive the object the more it distorts or curves space. It could be compared to how a bowling ball would bend a rubber membrane. He proposed that such a distortion of spacetime near a massive object like a star would also bend light.

Figure showing how a massive object like the Earth bends spacetime.

When Einstein first published his paper on general relativity in 1915, saying that space could be bent by a massive object, he was not taken seriously by most of his peers. However, in May 1919 a team led by the British astronomer Arthur Stanley Eddington photographed stars close to the sun during a solar eclipse on Principe Island located near the Equator in the Atlantic Ocean. Eddington found that starlight from stars close to the sun was bent by the sun just as Einstein predicted. This astounding confirmation of his theory guaranteed Einstein's global renown.

According to the theory, an extremely massive object could curve spacetime so strongly that nothing—no particle or even electromagnetic radiation such as light—could escape from it. Today we call such objects black holes, and astronomers have identified thousands of them. In the same way that mass pulls on space, Einstein's equations indicated that gravity can "pull" on time causing time to slow down. Hence, near the boundary of a black hole—the so called "event horizon"—the flow of time literally ceases.

The New Model for Energy, Matter, Space, and Time

Relativity theory presented a radical departure from the classical physics of Newton. But it was completely deterministic and never questioned cause and effect or suggested that observation had any effect on how bodies behave in spacetime. Nonetheless, it has caused scientists to modify their fundamental understanding of reality. Some of these modifications are counterintuitive and seem to go against common sense.

Probably the single most mind-boggling ramification of the theory is the connection between space and time. Because both of these are relative and intertwined, the classical concept of time as a constant everywhere in the universe must be thrown out. Only events in the four-dimensional continuum of spacetime have meaning. For example, a star goes supernova in our neighboring Andromeda Galaxy. This event is considered a point in the four-dimensional matrix of spacetime, but it will be observed on Earth 2.5 million years later—the time light takes to travel the distance. In other words, the spacetime separation between the event in Andromeda and our witnessing it on Earth is 2.5 million years. However, for the stream of light particles (photons) carrying information of the event, time and distance will be meaningless because during their transit, their internal clock is stopped as they squeeze space to a point. Hence, from the perspective of the photon, the supernova and its observation anywhere in the universe occur simultaneously. Within four-dimensional spacetime, the exploding of the star is a singular event.

According to relativity theory when an object moves through space some of its motion is diverted into time. When an object is not moving relative to another, then it is moving in time alone. If an object is moving, space will shrink and time will dilate (slow). For example, in the distant future, human beings might need to move to another planet with a suitable environment as the sun runs out of hydrogen fuel, begins to grow very large, and threatens to engulf the Earth. Next, suppose by this time humans have developed the technology to build a spaceship that can travel at 90 percent of the speed of light (270,000 km/sec). Astronauts set out for an inhabitable exoplanet orbiting a red dwarf star 20 light years distant. We might calculate that it would take them 22 years to reach the planet at that speed. However, after the astronauts reach full speed, they will calculate their distance to the star to be only 10 light years (because of the compression of space). They will then calculate that it should take

11 years to reach their destination. However, because their clock runs at half the speed as clocks on Earth, the event of their arrival after eleven years of their time will correspond precisely with their expected arrival by people on Earth (22 years). The astronauts and the earthlings do not agree on the time or the distance it takes to travel to the planet but the four-dimensional separation of the events corresponding to their launch and arrival on the new planet is something they can agree upon.

These two examples illustrate that only events in the four-dimensional continuum of spacetime have any meaning. Different observers that are moving relative to one another may not agree about time and distance, but they can agree on events. Events are singular points in the spacetime continuum—like seeds in a watermelon, they are positioned in fixed locations. Using this analogy, the Big Bang would correspond to the stem of the watermelon and time would progress on the horizontal axis. For any specific event (seed), the "now" would be one of the possible planes cut into the watermelon that went through the seed. Depending on the angle of the cut, we might experience the event in a different *now* depending on our relative position and speed, but the event would be fixed within the matrix of the watermelon.

In this analogy using a watermelon, a slice of the infinitely long "melon of spacetime" represents a particular now, while the seeds represent events and the stem corresponds to the Big

Bang. The angle of the slice depends on our relative velocity and influences our experience of "now."

What this means is that all events that we think of having happened in the past or will happen in the future are there in four-dimensional spacetime. This is the new picture of space and time demanded by Einstein's theory of relativity. Cosmologists call it "block time." The full matrix of spacetime includes *all* of space and time.

We are conditioned to think that at any given moment of time all space exists. At this very moment you are reading this, your hometown exists along with the planet Earth, the moon, the planet Mars, the sun, our Milky Way galaxy, and other distant galaxies. For the same reason the block universe of relativity requires that at any given point in space all time exists. This means that if we were endowed with four-dimensional sight, we could experience reality quite differently. Sitting at a fixed position in space—say our back porch—we could simultaneously witness all the events that took place there. We might watch dinosaurs walking by, and concurrently observe astronauts taking off for a distant exoplanet a hundred thousand years in the future. Since we do not have four-dimensional sight and since we are conditioned to experience the three dimensions of space and the constant flow of time, this new picture of the universe is very difficult for us to comprehend. However, we cannot get around the implications of relativity.

According to relativity theory, all past and future events are already "set" in four-dimensional spacetime. We perceive things changing in time, but this perception is illusory since from the perspective of the totality of spacetime, all events that have or will happen are already there and do not change in time. The past, present, and future are all equally real and the flow of time is something we create as a convenient way to help us cope with our three-dimensional experience of reality. In other words, our mundane experience of the flow of time is different from the time of spacetime. What we call "time" is our conscious experience of a constantly changing sequence of events. It can be called "conscious time" (physicists like to call it evolution time), and represents the conscious movement along one of the many possible timelines intersecting events in the wholeness of spacetime.

Einstein put it this way, "People like us, who believe in physics, know that the distinction between past, present, and future is only a stubbornly persistent illusion."[36]

One way to picture this is to consider a series of still pictures that are rapidly displayed on a cinema screen. The series of unchanging snapshots of the scene creates the illusion of movement, and the flow of time is associated with such movement. Yet, this perceived movement is actually nothing but the dance of three-dimensional snapshots of our world as it moves at a constant rate through the fixed four-dimensional matrix of spacetime. Events appear to unfold before us in time when they are actually "fixed" in the spacetime continuum. This implies that cosmic events are in a sense predetermined. However, if we believe in the idea that humans possess the capacity for choice or free will then on the scale of human activity our individual future would not be predetermined.

A summary of this new vision of time and space is outlined below.

- *All motion is relative.* There is no special frame of reference that can be considered stationary. As a result, space and time are observer dependent. Time and length may expand or shrink depending on the relative state of motion of the observer and the observed. According to relativity theory when an object moves through space some of its motion is diverted into time. As space shrinks, time slows, but the combined speed of an object's motion through space and its motion through time always remains equal to the speed of light. Space is transformed into time and time into space. This is the hallmark of a four-dimensional substance in which each dimension has both spatial and temporal aspects that are fully integrated and inseparable.
- *Movement causes space to divert to time.* When an object is moving slowly, then for practical purposes it is moving in time alone. If an object is moving near the speed of light, then it is moving mostly through space, and time for that object will be slowed significantly.
- *Four-dimensional spacetime does not change in time and is characterized by wholeness.* From the three-dimensional perspective of human experience, everything changes in time, but underneath our perception of physical reality lies the four-dimensional realm of spacetime in which all events are set in fixed coordinates of time and space.
- *A massive object distorts spacetime.* This bending of spacetime corresponds to gravity, and the distortion pulls on time as well, and it is slowed down. If astronauts were to orbit a black hole with its massive gravitational pull, their clock would slow significantly

relative to clocks on Earth while they were subject to the extreme gravity of the black hole. It is conceivable that after a few hours of "slowed time" near the black hole they could return to Earth and be younger than their grandchildren are.
- *There is an absolute speed limit for any element of physical reality—the speed of light.* Without such a speed limit, physical reality as we know it would cease to exist as objects could conceivably travel into the past creating impossible paradoxes.
- *Spatial dimensions are compressed at high speed.* The shape of an object such as a starship would look compressed or flattened to someone observing it as it passed by Earth at 90 percent of the speed of light. To the astronauts on the starship, everything would look perfectly normal since everything including their measuring devices would have shrunk the same relative amount. This is termed the Lorentz–FitzGerald contraction, named after Hendrik Lorentz and George FitzGerald, who developed an equation to describe the effect.
- *The now is not the same for observers moving relative to one another.* For example, if astronauts were traveling away from Earth at high speed their experience of now would be of events that already occurred on Earth, while if they were moving toward Earth they would experience events that have not yet taken place on Earth. The now experienced by the astronauts is just as valid as that of earthlings. Hence, the now, like the past and future, is observer dependent and mutable.
- *If we gained a four-dimensional perception of spacetime, then we might describe the experience as timeless or as an "eternal now."* Interestingly, this is how people have described mystical experiences.

Einstein's theories stand as one of the most important scientific discoveries of the twentieth century. Many of today's technological advances (e.g. the GPS system) depend on the relative mechanics derived from the theory. Furthermore, Einstein's theories of relativity have been verified by numerous scientific experiments and observations. Predictions made by the theory have proven correct and highly accurate. A recent example of this was the confirmation of gravitational waves that Einstein predicted. The miniscule distortion of spacetime that was produced by the collision of two black holes in a distant galaxy was picked up by the pair of LIGO observatories in Hanford, WA and Livingston, LA in 2016.[37]

Relativity has become an integral part of our understanding of how the universe formed and how it evolved from a dimensionless point. Implications of the theory have far-reaching ramifications on how we should describe reality. However, this description is nothing like the quantum mechanical description of reality. Relativity theory describes a physical reality that is continuous, local, deterministic, and with an arrow of time that begins with the Big Bang. It has no fuzziness and, it would seem to be incompatible with quantum physics.

The reason for the disparity of these two great theories is that they describe the working and motions of reality at *different hierarchical levels*. Relativity theory only touches on what we would call "objective physical reality." Spacetime is a subtle element of this reality, but nonetheless spacetime represents the "container" for physical reality. Everything we associate with physical reality including our bodies lies within spacetime. Distance and time are fundamental aspects of this domain of reality, and everything that occurs within spacetime is continuous, deterministic, and local.

On the other hand, quantum mechanics reveals the existence of a subtler level of reality, which is characterized by the domain of the wave function. As we have seen, this domain is timeless, nonlocal, and whole. Information about the quantum state of one particle can be transmitted to another particle instantaneously even over cosmic distances. Theoretically, everything in the domain of objective physical reality must be an emergent property of this domain—including spacetime itself.

Although physicists have been unable to develop a theory that connects quantum physics and relativity, it is a topic of active investigation. The most promising avenue in this regard is called quantum field theory. Physicists in this area of research are seeking to develop a mathematical description of how spacetime can be understood to emerge from a quantum field. The math works to describe a quantum field that obeys Einstein's equations of general relativity for an object with moderate mass like our sun, but the theory breaks down when gravity is strong as in black holes. However, there is real hope that sometime in the future gravity may be explained by quantum mechanics. Thus it is reasonable to think that someday physicists might develop a concise theory that shows how spacetime emerges from a wave function. They are not there yet, but it is possible that, like matter, spacetime will be shown to be an emergent phenomenon and therefore not fundamental. Such a scenario in which all the elements of the universe can be thought of as emerging from the subtle domain of the wave function is entirely consistent with the spiritual worldview.

It is obvious to anyone that the "time" in Einstein's spacetime is not the same thing as what we call time. Absent conscious awareness, there would be only the unchanging wholeness of spacetime. *Consciousness is the missing ingredient that produces experiential reality.* Lacking consciousness physics has no flow of time. Hence, it defines time as illusory, but for us time is not an illusion. We inevitably age from the time of our birth, grow old, and die. This is merely another illustration of the importance that consciousness plays in any understanding of reality, and for physics to ignore this indispensable role that consciousness plays in any theory of reality is illogical.

However, today few scientists seem willing to consider or talk about the implications that relativity theory has on the materialist worldview. This is probably because to do so would be to admit that our experience of a time-flowing, three-dimensional world is merely a shadow of a more encompassing realm of wholeness that includes all of space and time. This is the same picture of reality painted by idealist philosophers such as Plato, and it would be contrary to their cherished materialist view of reality.

PART TWO

WHY MIND CANNOT BE REDUCED TO BRAIN

So far, we have talked a lot about the materialist worldview of reality and why it has changed since the discovery of quantum mechanics and relativity theory. It can be termed a "bottom-up" ontology because it begins with matter (and its associated energy and fields) and from there postulates that living organisms evolved that eventually display mind and consciousness. Modern-day materialism now accepts the possibility that there might be other subtler levels of reality, such as the one described by the wave function, that do not follow the usual laws of physics. This is OK. But most are loath to consider the possibility that mind and consciousness are anything but the result of the interactions of physical matter.

On the other hand, spirituality has an alternate description of reality. We call it a "top-down" ontology, because it postulates that consciousness is the fundamental "stuff" from which the cruder aspects of reality are derived. According to this worldview, mind and matter are epiphenomenal to consciousness. Even though this worldview is consistent with that of most religions, spirituality is not religion, but it does appear to be the foundation for all the world's great religions since their founders appear to have experienced a spiritual rebirth or a mystical understanding of the oneness of creation.[38]

We might ask the question whether either of these two ontologies is falsifiable. I believe the answer is yes. If science could prove that mind is *nonphysical*, then materialism would be shown to be false. Secondly, if it were proven that mental communication is possible in ways that do not use the physical senses this would disprove the theory. Finally, if it were shown that consciousness is nonlocal or that there is a "universal mind," then the materialist paradigm would be proven false. Key postulates of materialism hinge on the assumption that mind and consciousness are epiphenomena of brain, and therefore derived from matter and thus local faculties.

Similarly, the spiritual worldview would be proven false if it could be shown that consciousness can be produced by physical means. For example, suppose computer scientists were able to construct advanced artificial intelligence (AI) that possessed conscious self-awareness. According to the model of reality postulated by spirituality, consciousness is the foundation from which the material world is created. It would be impossible for it to be reproduced through physical processes. Currently, neuroscience is a long way from even forming a theory of how consciousness is created by the brain, and computer scientists are nowhere near developing AI having conscious awareness—so this is not going to happen anytime soon. There is also the problem that even if AI was developed that from all outward appearances had "machine consciousness," it might be impossible to prove that it was actually self-aware, since we cannot get into the mind of a machine—or any mind for that matter.[39]

Of course, there is also the question whether science can prove anything. Scientific proof is nothing like a mathematical proof. It is normally defined as *overwhelming evidence, so conclusive that it would be logically or probabilistically unreasonable to deny the proposition*. This may be as close as we can come to scientific proof, since even if there is a pile of evidence for a theory, it is conceivable that some new evidence might come along that forces us to throw out the theory or at least modify it.

Probably the best we can hope for in our analysis of materialism vs. spirituality is to weigh the evidence for the two theories and decide which fits the data better. In the end, the data should be the final arbitrator of which theory of reality is correct. If the evidence falsifies a theory then according to science that theory must be abandoned or at least modified to fit the data. One approach to weighing the evidence is to go back to the method proposed by Reverend Thomas Bayes or "Bayes's Theorem." It says that we might start with a 50 percent probability that either theory of reality is correct, and then build up the probability that one is correct vs. the other based on the evidence. In other words, we should consider which of the two theories offers the best explanation for the evidence.

We might start this process by weighing the evidence for whether mind can be reduced to brain. However, before we get into the question of which ontology is more credible it is prudent to ask the question—what can we really know about reality?

7

HOW CAN REALITY BE EXPERIENCED?

Sensing Reality—Does the Brain do it All?

THE MATERIALIST WORLDVIEW POSTULATES that the physical world that we experience with our sense organs is real. This makes perfect sense since other people, not to mention animals, appear to experience the same reality that we experience. We take for granted that normal human beings experience the outside world the same way we do. When they describe the color of a rose as red, the smell as sweet, and the petals as silky smooth, we assume that their experience is the same as ours. However, considering what we know about how our sense organs work, we would conclude that our experience of the rose is generated by our brain.

Take for example the color of the rose. Light consisting of a spectrum of many wavelengths hits the petals but most of the light is absorbed and only the longer wavelengths corresponding to the color red are reflected. This light enters the eye and stimulates certain red-type photoreceptor cells (cones) in the retina. In these cells, a chemical transformation takes place that causes electrical signals to be generated that are carried to the visual cortex in the brain via the optic nerve. The actual image of the rose with its color is experienced in the brain. We would be correct in saying that the sense organ for sight is the region or regions in the brain that are responsible for generating the image (form) of the rose along with its color. The eye (along with the nerves communicating the information to the brain) is only a gateway organ for sight. The other senses of hearing, touch, taste, and smell work analogously.

Even the idea that there exists something like a screen that displays the image of the rose in our brain is questionable. Neuroscientists today know that visual stimuli, which include color, form, and motion, are handled by different parts of the brain. They are almost unanimously opposed to the idea that there is some nonphysical witnessing entity (a self) that observes those parts of the brain that are responsible for our sensory experiences.

Hence, our experience of reality is constructed by our brain resulting from the nerve signals it receives plus our prior conditioning. For example, color is not a property of an object but arises from our interpretation of signals received by the organ of sight (visual cortex) in the brain. We were taught to identify the word red with the experience of light of a particular wavelength. This goes for all our five sense organs and applies generally to all our experiences of the physical world.

We have no way of experiencing physical reality directly. The best our brain can do is create a representation of that reality based on our prior experience. We might be outside enjoying a cup of coffee while reading the newspaper. No light enters our brain or the sound from the flapping of a bird's wings. The roughness of the paper, the sweetness of the added sugar, or the smells of the freshly brewed coffee are interpretations of nerve impulses that have been transmitted to our sense organs in the brain.

This part of the story is consistent with the materialist model of reality. There might be a small problem explaining how our brain is able to create a unified experience of reality given the fact that our sense organs are in anatomically separate parts of the brain. However, neuroscientists have theories about this.

We might also ask the question of exactly *who* does the experiencing and interpreting of the electrical impulses generated by the gateway organs. According to the materialist model, the brain itself is generating a representation of physical reality existing outside our body and also interpreting and experiencing this reality. There is no reason to postulate the existence of a nonphysical entity—self, mind, or consciousness being involved in the process. However, in order for the sensory data that is both arriving and stored in the brain to be processed, something in the brain has to look at it. This something could be termed the *I* or *self*. This self would need to be generated by the brain. In other words, we need to postulate that there is built into our network of brain neurons the capacity to simulate the experience of self-awareness—a part of our brain capable of witnessing other processing parts of the brain such as the sense organs. This brain-generated self could be compared to a little person (homunculus) that is within us who has its own system of memory and awareness along with the skills and memories needed for processing the data. This "little self" would need to embody *all* the capabilities we wish to explain in the first place.

Can We *Know* Physical Reality?

Everything we know about the world comes to us via our sense organs and is thus *experiential*. Science relies on experiment and experiments rely on experiences. However, all experience is subjective—brain created. The evidence for the existence of a physical reality that is separate from us can only be based on subjective experience. Other people describe the same world we experience and that gives us confidence that our experience is of a real world. In addition, our bodies constantly remind us that we have a physical part of our being that we must attend to. Isn't this proof that there is an objective reality out there? The answer is no, because these experiences including our bodily sensations such as pain and pleasure come to us through our brain. They are subjective experiences that we naturally project onto an objective world and a sense of body that is created by our brain.

Surely, the scientific method can be used to prove the existence of an objective reality. Science is devoted to the study of the structure and behavior of the physical and natural world through observation and experiment. The only problem is that those observations and experiments rely on subjective experience.

Suppose we want to test the scientific hypothesis that there is an external reality out there that is separate from us. For example, we might find a piece of quartz that appears to have some gold imbedded within it. We have the rock analyzed by a chemist who examines it under a magnifying glass, weighs it, crushes it, and performs a chemical assay. The analysis determines that the rock indeed contains gold and that it comprises a certain percentage of the rock's mass. It does not matter which chemist does this examination, if they use the same methodology, they will probably report similar results.

Sound scientific methods were employed in this analysis, but every element of the examination depended on subjective experience. Unfortunately, such experiments cannot demonstrate that the rock constitutes an objective entity that is separate from us. The only measuring stick was an internal or subjective experience of some elements of physical reality. We can claim that our experience depends on the reproducible and consistent input from our sense organs and from the reports of others that are similar to our perceptions. This is fine, but it cannot *prove* that such an external reality exists. Just like information from our senses, science does not provide conclusive evidence in this case—it would be circular reasoning. The experimental information that informs us about the physical world is a creation of the physical world (the brain). Science can provide us with data about how things *relate* to one another, but it cannot tell us what their *fundamental* nature is.

The question of what we can truly know about reality was also asked by Carl Jung:

> How in the world do people know that the only reality is the physical atom, when this cannot even be proved to exist at all except by means of the psyche? If there is anything that can be described as primary, it must surely be the psyche and not the atom, which, like everything else in our experience, is presented to us directly only as a psychic model or image.[40]

John Wheeler wrote, "All this talk about an external world—that's what's theory!"[41] While Erwin Schrödinger expressed the same thought when he said that based on the evidence of the quantum, the actuality of the objective world remains a hypothesis. Or as Werner Heisenberg put it, "The idea of an objective real world whose smallest parts exist objectively in the same sense as stones or trees exist, independently of whether or not we observe them—impossible."[42]

In other words, our knowledge of the physical world is created by our brain and hence we cannot know with certainty that there exists a physical reality separate from us. Therefore, it is possible that what we perceive as external reality is nothing but a creation of our brain. The idea that only our mind is sure to exist and that all reality is created by our mind is known as solipsism. I am not promoting this philosophical idea, just pointing out that there is no way to know whether there is an objective reality existing separate from us.

What Can We Know Then?

If we cannot know that physical reality actually exists out there, what can we know with certainty? Perhaps we can fall back on the reasoning of Descartes, "I think therefore I am." However, this statement is merely a truism or tautology based on the words used. It can be reduced to, "I am, and therefore I think." What Descartes was really trying to say was that since I am conscious, I must exist. But even this statement might be questioned because it implies that when I am unconscious, as in deep sleep, I may cease to exist. So if our consciousness is something that can be turned on and off and be described differently when we are awake or dreaming or under the influence of drugs, is there still a part of our being that is always there and unchanging? The answer is yes, and we call it the *self*. The self is our sense of identity and it has continuity. It is our sense of "I am." It has not changed since we were a child even though our physical body has undergone radical changes. The self links all our experiences together in a consistent pattern. It is immediately present after a bout of unconsciousness reminding us of who we are now, who we were yesterday, and what we would like to do tomorrow. Our self-identity is a permanent, continuous aspect of our being and the foundation upon which all our experience is based. Therefore, if there is anything we know with certainty it must be our self-awareness. However, it is hard to see how the self could be a physical aspect of the body or brain because these undergo continuous change. The self does not change—we always identify as the same person.

The self is linked inexorably to our awareness of being a singular being. It could also be called our unit consciousness or "I feeling." We are conscious of the external world and conscious that we are witnessing that world. That is, we are self-aware. We could say that the self has the capacity to observe itself as well as the stream of information produced by the brain.

The Experience of Separateness

We have seen that both quantum physics and relativity theory lead inexorably to the conclusion that the universe is an indivisible whole. If this is so, why do we experience separateness? David Bohm suggests the answer is due to our need to make order out of our perceptual experiences. He argued that this has led to the illusion that what is real is an objective reality consisting of individual objects separate from us. He argued that this materialist worldview has permeated both science and most people's way of thinking. We might ask the question: is there any scientific analogy to thinking in terms of wholeness rather than separate parts. Bohm suggested that we compare instruments such as the microscope or telescope that function to amplify our separateness to a hologram in which every part reproduces a whole image. As he would put it, the image is enfolded in every part of the holographic material.

The actual experience of wholeness would appear to be rarely attained. Mystics describe such experiences, and sometimes such experiences occur spontaneously. Others claim to experience wholeness through persistent spiritual practices. Wholeness is a postulate of the spiritual worldview, which argues that this experience occurs upon merging our unit mind with cosmic mind or consciousness, and that this is the ultimate goal of human existence. In Chapter 11, we will learn more about mystical experiences.

All this philosophizing about reality leads inexorably to the conclusion that the only thing we *know with certainty* about reality is our own existence. Does this mean that physical reality cannot be absolute reality? The answer is no. All we can say is that there appears to be a "relative reality" out there that sane people agree exists. But it is something we can only experience subjectively. Whether we choose to consider it "absolute reality" does not really matter—we have no choice but to deal with it. On the other hand, our inner world governed by the self is the witness of experience and definitely exists. Whether the self is a product of the physical world is open to question, and the truth of such a proposition should be decided by weighing the available evidence.

8

THE MATERIALIST MODEL EQUATING MIND AND BRAIN

MUCH OF THE RESEARCH by neuroscientists into brain anatomy and function in the last few decades has reinforced the doctrine of a physical basis for mind. New methodologies based on behavioral, clinical, pharmacologic, genetic, neurosurgical, and electrophysiological probes as well as neuroimaging techniques, such as functional MRI, have increasingly demonstrated the close linkage between brain physiology and mental states. This has convinced most neuroscientists that almost all mental states have a physiological basis and that there is no longer any need to consider the dualistic separation of mind and brain. They also point to the fact that damage or changes in the brain cause changes in mental functioning and consciousness.

Messages are carried down the length (axon) of a nerve cell (neuron) electrochemically. The electrical activity of the brain may be measured by an electroencephalograph (EEG). The EEG measures voltage fluctuations resulting from ionic currents within the neurons using electrodes placed on the scalp. Neurons talk to one another by releasing chemical messengers—neurotransmitters—across the junction between them. And when charged particles move they create a magnetic field that can be picked up outside the skull by instruments such as a magnetoencephalograph (MEG). Using these techniques, neuroscientists can map out thought processes. It is possible that any thought we have can be associated with tiny electric circuits and the motion of chemical particles in our brain. This means that thoughts may be reduced to specific electrochemical events occurring in specific neurons in our brain.

The modern theory of how the brain stores and encodes information is the network model. A typical human brain weighs about three pounds but contains a trillion cells, 10 percent of which are nerve cells; hence they number about one hundred billion. On the average, each neuron can receive signals and be connected to about five thousand other neurons. This means that the total number of possible connections (synapses) for the brain's neuro network is about a quadrillion (10^{15}). According to the network model, memories and learning are distributed across many neurons and their connections or pathways, resulting in a network that continues to function even when parts are damaged or removed. The model describes the brain as functioning like a super-duper computer, and like a computer with tons of memory chips, personal memories and knowledge are "stored" in the neuro network much like files on a computer.

Neuroscientists have proposed that the brain network could be comprehensively mapped in a way similar to the human genome. They call this the "connectome." The human connectome would include the mapping of all neural connections within the nervous system of a healthy individual.

Currently the effort to build a database for the brain's neural connections is the focus of the Human Connectome Project, sponsored by the National Institutes of Health (NIH). Having such a detailed map of the brain might allow neuroscientists to map all the thoughts, associations, memories, and ideas of a human being. This would appear to be a very difficult undertaking and currently researchers have been successful in constructing the full connectome of only one animal: the tiny roundworm *C. elegans*.

Memory

Research has shown that memories are not stored in any specific location of the brain but are spread throughout the brain much like a hologram stores a three-dimensional picture throughout its entire matrix. Memories may consist of exquisite details and mental pictures seen from a third person point of view, and such detailed recollections may be brought out by electrical stimulation of the brain or hypnosis.

It is abundantly clear that damage to the brain can have an effect on memory. One example that comes to mind is the devastating effect that

Alzheimer's disease has on memory. In addition, damages to specific structures of the brain have been shown to affect memory.

As a volunteer in the adaptive program at our local ski resort, I had firsthand experience of how damage to the brain could affect short-term memory. My client (DW) had suffered cardiac arrest at the age of 45, but was resuscitated. The resulting stoppage of blood flow to his brain for a few minutes apparently damaged parts of his brain that are important for processing short-term memories. In particular, the hippocampus is believed to be responsible for learning new information and its neurons are highly sensitive to oxygen deprivation. DW was 58 when I began serving as his ski guide. He loved to ski—once being on the professional ski-racing circuit—but he had lost the ability to store current information and had no idea where any of the ski runs would lead him. We could have skied a single run all day and each experience would be new to him, but obviously I would be bored doing this and chose what I considered a more interesting program. DW did not recognize or remember my name even though we skied together weekly. He had no problem remembering details of his life before his heart attack, but damage to his brain did not allow current events to be stored in his long-term memory. For example, when I asked DW the name of his dog, that he walked daily, he said it was named Max—when in fact she was named Crystal. Max was the name of his dog from before his heart attack. DW could operate pretty well in the here-and-now. It is as though the camera recording his life worked well, but its playback function was defective.

Amnesia is a term used to describe a deficit in memory. It is often temporary and may be caused by brain damage, disease, or by various drugs such as midazolam (Versed). Examples like these of brain-damaged individuals, disease states like Alzheimer's, and drugs that induce memory loss have given neuroscientists confidence that memories are stored in the brain and there is no need to appeal to a dualistic explanation like that of Descartes that calls for a nonmaterial mind interacting with a physical body.

Can Functional MRI Tell What we are Thinking?

Functional magnetic resonance imaging (fMRI) measures brain activity by detecting changes in blood flow in various parts of the brain. The technique relies on the fact that cerebral blood flow and neuronal activation are coupled. When an area of the brain is in use, blood flow to that region increases.

By comparing the fMRI scans of people's brain after viewing certain images neuroscientists can create a detailed map of how they react to specific stimuli. Depending on the type of image shown—for example gruesome, beautiful, sexy, etc.—they see different parts of the brain light up and conclude that this shows that our sensory perceptions are encoded in these locations. They can even tell whether we are watching a video of someone enjoying a glass of wine or crying over the loss of a loved one. The data clearly shows that there are patterns of nerve activity in the brain that mirror our observed behavior and our inner experiences. Such studies contribute to the idea that eventually all subjective experiences will be explained by physical phenomena occurring in the brain. The only problem is that after several decades of research, a consistent pattern for mapping subjective experiences with measurable brain processes has failed to emerge. But neuroscientists are encouraged by such studies and think that it is only a matter of time (and money) before a one-to-one mapping is accomplished. Some are even thinking that it is time for them to tackle the biggest problem of all—explaining consciousness.

The Brain and Consciousness

Neuroscientists consider a brain basis or explanation for consciousness to be the "hard problem." It is easy to explain the difference between being awake and asleep or conscious vs. unconscious. On the other hand, it is quite difficult to explain why sentient or self-aware organisms have qualia or phenomenal experiences. Why is it that we have qualitative experiences as the brain engages in information processing from the sense organs? For example, our visual cortex is stimulated by light and our experience is one of elation as we contemplate a beautiful sunset. Alternatively, our auditory center in the brain is stimulated by the sounds of an orchestra and we feel great delight in the music. How can we explain how something like a mental image can evoke emotions of love, hate, or fear? How can the physical processing by the brain give rise to emotion and a rich inner life? In other words, the hard problem is to come up with an explanation for how subjective experience and a sense of self can result from purely physical processes going on in the brain.

The man who first came up with the term, "hard problem," was Australian philosopher and cognitive scientist David Chalmers. He put it this way, "Why does the feeling which accompanies awareness of sensory information exist at all?"[43] He argued that the easy problem involved cognition, i.e. why a rose appears red and could at least theoretically be answerable via the dominant strategy in the philosophy of neuroscience—physicalism. On the other hand, the hard problem of explaining subjective experiences has what he called an "explanatory gap" from the objective to the subjective. He argued that physicalist explanations of mental experience will always fail because mental states are ontologically distinct from and not reducible to physical systems. In other words, we lack a bridge from mindless particles to mindful experience. Framed another way the question is: how exactly is it possible for a certain highly complex pattern of atoms interacting according to the laws of physics to become self-aware and have subjective experiences?

Notwithstanding the difficulty of the problem, researchers in neuroscience and psychology have learned a lot about consciousness. The best current theory is that it is not a single thing or function. It may be a conglomeration of things like self-awareness, wakefulness, receiving and responding to sensory inputs, memories, subjective experiences, intention, thinking about the past and the future, and imagination. Neuroscience is clearly in its infancy in developing a comprehensive picture of how the brain creates consciousness, but much progress has been made and many neuroscientists believe that a complete explanation will be forthcoming in the future.

There are scientists that believe that one day AI will advance to the point that intelligent machines will speak, reason, solve problems, be introspective, and for all practical purposes become sentient. Some researchers believe that the global neuronal workspace (GNW) theory of consciousness might explain how this could be accomplished. GNW postulates that architectural features of the brain's neuro network are what give rise to consciousness. Conscious states are produced by the way the workspace algorithm processes the relevant sensory inputs, motor outputs, and internal variables as they relate to memory, motivation, and expectation.[44]

There is an alternate path that researchers in AI are investigating—integrated information theory (IIT). IIT attempts to explain what consciousness is and how it might be associated with certain physical systems. Given any such system, the theory predicts whether that system is conscious and to what degree it is conscious. It even tries to predict what particular experience it is having. According to IIT, a system's consciousness is determined by its causal properties and is therefore an intrinsic, fundamental property of a physical system. Scientists working on machine consciousness anticipate that it is only a matter of time before it will be attained. All it might take is a computer that can simulate the neuro network of the brain—perhaps a quantum computer. Of course, there are major hurdles to overcome before sentient machines can be built. Currently the most advanced computer networks have about 10^9 connections and the brain has 10^{15} or more. And while it might be easy to design a computer program that could translate this paragraph into Japanese, it is quite another matter to design one that could understand its meaning, and even more difficult to also have the program "feel" satisfaction once this was accomplished. In fact, John Searle pointed out that a computer cannot process meaning because it only has the capability of processing symbols. Any attempt to process meaning would require "meaning symbols." Then new symbols would be needed to process those meaning symbols—ad infinitum.[45] According to Searle, no computer will ever be able to process meaning or be considered conscious.

The Self and Unity of Experience

As mentioned earlier, there appears to be an aspect of our being that does not change over the course of our lifetime. Even though the atoms that make up our body may turnover almost completely in the course of a year, the pattern that is *us* does not change. We call this pattern the self. The self must be a higher-level pattern than the arrangement of atoms in our body since these arrangements constantly change, and there is almost nothing of us physically that is the same now as when we were a child. According to the current model of neuroscience, the self is created by brain from its collection of stored information and memories. It is merely a subjective feeling like love.

Along with the sense of self-hood, we have a *unified* experience of reality. It is now known that sensory inputs are handled by different mechanisms and/or anatomically separate parts of the brain. For example, when we observe a palm tree gently swaying in the wind the visual stimuli include color, form, and motion. We know from brain scans that these are handled by different parts of the brain; yet somehow we have a unified visual experience. Since there is unity of experience there would either have to be what is termed "anatomical convergence" or another mechanism. Anatomical convergence means a place in the brain where everything comes together to help create the unified experience of the palm tree. However, no such convergence exists in the brain, and therefore the unity of self-awareness and experience must be achieved by some other means than brain anatomy. Today the prevailing theory of neuroscience for how separate regions of the brain are "bound" to one another involves large-scale gamma-band oscillatory electrical activity.[46] Widely separated neural populations can be synchronized by these oscillations providing an explanation for why there is unity of conscious experience.

Free Will

We seem to influence the unfolding of reality through our actions, and we seem to be able to freely think thoughts and make decisions based on our desires and inclinations. Yet, according to the physicalist theory, we are nothing but patterns of particles that obey physical law. At the most fundamental level—the particles that make us up—are responsible for our thoughts and behaviors, and result from the complex shifting of these particles. Our rich spectrum of behavioral responses is due to our extremely complex internal organization, and this produces the illusion of autonomous action. However, we are governed only by physical laws, and the laws of physics cannot be violated. Hence, there is actually no volition allowed in how reality will manifest.

Furthermore, the state of our particles today is fully dependent on how they were yesterday. In fact, since we have no way to alter the physical particles that make us up, we could theoretically trace everything back to the Big Bang, which was the original source of all particles. Knowing our past condition completely we could predict with certainty how we would react in the future.

This analysis of free will is the logical extension of the materialist worldview. It might seem rather stark, but it is a direct result of applying the principles of this worldview to a human faculty that we take for granted. It is a simple extension of the principle that what applies on the microscopic scale—i.e. no freedom of choice for particles—must also apply on the macroscopic scale. We might object and point out that quantum mechanics provides only probabilities—not certainties. With the possibility that *chance* could play a role in how things evolve, there might still be room for choice. The problem is that the Schrödinger equation governs the mathematics of how things evolve, and it is completely deterministic. Although quantum mechanics says there might be many possible future outcomes, these are still set in stone and there is no room for free will.

Here is what a leading proponent of the materialist worldview, physicist Brian Greene, says about free will:

> We recognize that although the sensation of free will is real, the capacity to exert free will—the capacity for the human mind to transcend the laws that control physical progression—is not.[47]

On the other hand, another proponent of materialism[48], Sean Carroll, leaves some room for a modicum of free will. He writes:

> In a world governed by impersonal laws, one can argue that there is no such ability (free will). Given the quantum state of the elementary particles that make up me and my environment, the future is governed by the laws of physics. But in the real world, we are not given that quantum state. We have incomplete information; we know about the rough configuration of our bodies and we have some idea of our mental states. Given only that incomplete information—the information we actually have—it's completely conceivable that we could have acted differently.[49]

It is Carroll's argument that free will might still exist in a reality governed by physical laws in the sense that we are not privy to the quantum state of all particles. He also states that there might be alternative varieties of freedom—such as an almost unrestricted range of behaviors. According to Carroll, this would not violate the laws of physics.

The majority of neuroscientists today accept the physicalist model that mind and consciousness are merely products of brain activity—epiphenomenalism. After all, we evolved from primitive ancestors that gradually developed self-awareness. There is no reason to adopt a model which hypothesizes that mind is somehow a different, more transcendental, nonphysical entity that acts through the brain when every aspect of human mentality can be adequately explained by physical means.

However, we might ask the question whether neuroscience today is in a position similar to physics at the turn of the last century—willing to stick to a model despite the fact that there are a number of "anomalies" that are inconsistent with it. In the next few chapters, the discussion will turn to why these anomalies pose a real threat to the physicalist model of mind equals brain. However, just like most physicists who were convinced that the classical Newtonian view of reality was correct, most neuroscientists do not appear interested in investigating phenomena that might upset their preconceived notion of how nature works.

9

THE DIFFICULTIES IN EQUATING MIND AND BRAIN

MATERIALISM EQUATES MIND AND brain, but if David Chalmers' assessment of the role consciousness plays in subjective experience is correct, then this important aspect of mind is not explicable in terms of brain activity. This is because qualia are not by their nature reducible to physical systems. For example, do we really believe that we could create a computer that would feel love for its designer? The alternate hypothesis of spirituality does not have this problem. Here mind is thought to be a qualified aspect of consciousness and matter is a qualified aspect of mind. An analogy is how steam is transformed into liquid water, and water is transformed into ice. Each phase might outwardly appear as a different substance, yet they are just different forms of the same thing. Mind by its very nature as a qualified form of consciousness has the capability of having qualitative experiences.

An example of the difficulty in explaining qualia using the physicalist doctrine that mind can be equated with brain might be illustrated by an example. Suppose a person is born with anosmia—the inability to sense odors. Let's call her Jane. Normally humans have about 400 types of scent receptors and can distinguish up to a trillion different odors. Next, let's suppose that neuroscientists have developed a detailed map of the brain and identified the precise location of all the neurons that are responsible for the sense of smell and have even been able to determine the exact combination of neurons that are fired to produce all these different odors. Finally, let's suppose they could implant a chip in Jane's brain that has electrodes placed so that the specific neurons needed to produce any odor can be stimulated by a computer. After the surgery to implant the chip and other hardware connecting it to the computer, her doctors might test the implant by telling the computer to stimulate the select group, of say 500 neurons, that would create the scent of a rose. Jane would for the first time experience the sweet scent of a rose. No doubt, she would be thrilled to have her first direct experience of smell, and no doubt, it would create a very strong inner response in her mind. It might even evoke memories and emotions from her youth when her mother was tending roses in their garden when she was five.

As Jane is introduced to other odors, these would probably produce other feelings and experiences in her mind. She would immediately recognize some smells as pleasant and others as offensive. Jane would gain knowledge of new conscious experiences that accompany her brain's response to stimulation of its nerves. Even if neuroscientists had completely mapped out the brain's physical workings they would never be able to predict Jane's reactions or feelings that might be associated with each new smell she experiences. Knowing everything there is to know about the brain would still not explain her subjective experiences. We might conclude that even if we knew every detail of the brain's physical processes there is something missing and this missing piece is subjective experience.

Materialists like to say proponents of the "mind ≠ brain" camp are dualists. They say that this worldview is similar to the one proposed by Descartes—mind is a separate substance than physical matter. Such an analysis of mind and matter is incorrect. First, top-down ontology can be described as *monistic idealism*. Everything is formed from consciousness.[50] It is a holistic ideology. Secondly, this view is consistent with the fact that conscious experience is intrinsically holistic. All its features including its subjective point of view, conceptual understanding, unity of experience, and awareness of a singular self are holistic. Experience cannot be divided into different feelings about the physical and mental world—all feelings are mental. In fact, materialism is dualistic—or more accurately pluralistic. Everything is reducible to separate parts.

The question of whether physical reality exists or whether only our perception is real does not come up if we relegate physical reality to the arena of "relative reality," and consciousness to the arena of "absolute reality." We experience everything in the outside world through our sense organs and brain. This creates an external-internal dualism between, for example, an object such as a chair and our internal perception of it. What is real then, the chair or our perception of it with our mind? The answer is both. The chair represents relative reality and our perception of it ultimately depends on our conscious awareness of the chair, which is nonlocal and ultimately identical in quality—but not in quantity—to cosmic consciousness (absolute reality). Mind serves as the intermediate between the physical object, the brain, and the observer (consciousness or self). There really is no dualism—everything is composed of consciousness. But as human beings, we are fully immersed in the manifest universe. We are incapable of experiencing the non-duality of reality without shedding the illusion of separateness (ego), and this only occurs in cosmic union.

With the spiritual worldview, most of the difficulties in explaining qualia, self-awareness, and conscious experience evaporate. The so-called hard problem is no problem. However, it is fair to ask what evidence there is that mind cannot be reduced to brain. Certainly, the majority of experts in this field and most scientists today subscribe to the theory that mind is reducible to brain activity—and a strong argument for this was made in the last chapter. No doubt, it has been easier for neuroscientists to study how changes in the brain affect mental states. However, the mind-brain equivalence struggles to explain a number of mental functions and seems to break down when studies switch to how a change in a mental state (mind) affects a physical state (brain/body).

The Brain, Self-Awareness, Memories, and Intentionality

Clearly, there is a very strong correlation between mental states and the health and functioning of the brain. There is no doubt that damage to the brain can affect memory, sensory experience, conscious awareness, capacity to use language, and the ability to feel emotions. However, it is *not true that correlation is causation*. For example, we know that the music coming from our car radio is not produced by the radio. However, if a person from the seventeenth century saw and heard our radio they would be amazed, and probably assume that somehow an entire orchestra was contained in the device. In a similar way, the brain is required in order for the mind to gain expression but this does not mean that the brain is the cause of mind. The question is not whether the mind and brain are intimately connected but in the nature of the connection. Clearly, a mind that is not generated by the brain is still focused and constrained by the brain—a nonmaterial mind still requires a healthy brain to function in the body. But it is a mistake to equate the object that "plays" the music (brain) with that which creates it (mind).

A brain basis for self-awareness faces the problem that no structure or group of structures in the brain appear to generate self-awareness. It is true that drugs and injuries to certain areas of the brain will induce unconsciousness, but whenever there is even the faintest spark of consciousness, even in the most confabulated or dream state, there will be at least some identification with self. Clearly, if self-awareness is created by the brain, it must result from the totality or gestalt of brain activity.

We experience unity of self-awareness, by which we mean that our experience of reality is holistic. It is not one in which each sensation remains separate and distinct from each other. This unified experience of reality is difficult to explain using the theory that brain is the basis for mind and consciousness since sensory inputs are handled by different mechanisms and anatomically separate parts of the brain. Yet, we experience the outside world in its totality. Neuroscience has yet to come up with a viable explanation of why we have a unified conscious experience.

In addition, if self-awareness is generated by the brain, how is it possible that lucid consciousness with self-awareness and unity of experience is reported to continue during cardiac arrest and other stoppages of blood flow to the brain? This is observed in out-of-body and near-death experiences—cases where there may be no measurable electrophysiological brain activity. Such experiences, if real, would indicate that being a conscious, singular person is not adequately explained by the physicalist doctrine that reduces mind to functional brain activity. In Chapter 12, we will have the opportunity to learn more about these experiences.

The physicalist model of self-awareness postulates that self-awareness is a creation of brain activity. This means that the processing of sensory data that is both arriving and stored in the physical brain is looked at by another part of the brain that could be termed the *I* or *self*. As mentioned earlier, this self (homunculus) behaves like a little being within us that must have its own system of memory and awareness along with the skills and memories needed for processing the data. The same goes for the retrieval of memories. In order for memories to be looked at, we would need a second memory system that has the skills and memories needed for such retrieval. Hence, we need a second homunculus to look over the shoulder of the first one and so on.

One of the first theories for how memories are stored in the brain was the "trace" model. Experience causes physical changes in the brain, which can later be recalled when these physical traces or "engrams" are retraced cognitively. This model worked for simple creatures such as slugs that clearly undergo changes in their primitive nervous system in response to changes in their environment. However, such learned responses are nothing like human memory with its ability to recall at will details of past events (autobiographical memory) or general knowledge (semantic memory). In addition, neuroscientists have been unable to locate or identify any engrams in the brains of test animals.

Today researchers believe that memories are not stored in any specific location in the brain, but are stored in the brain's hierarchical network. This would explain how memories might be spread throughout the brain much like a hologram stores a three-dimensional picture. This is seen, for example, in experiments using rats that were trained to run a maze. When increasing amount of tissue was removed from their cerebral cortices before reintroducing them to the maze, their memory was degraded; but remarkably, it made no difference where in the brain the tissue was removed.[51]

Memories may consist of exquisite details and mental pictures that one is not even conscious of most of the time. These detailed recollections may sometimes be brought out by electrical stimulation of the brain or hypnosis. Most memories are not simple replays in the mind of the original event. Instead, a person will see himself or herself witnessing or taking part in an event. People are aware that they saw or did something. In other words, there is self-awareness, which is a mental function, not just a replaying of events. If memories were only stored physically in the brain, then it seems reasonable that they would appear as a replay and not witnessed from a third-person point of view.

Along these same lines is the question of how memory is placed in time. Somehow, the mind "places" the memory in a personal time-line in such a way that it relates to other dates, names, events, etc. Thus, instead of recalling a specific memory or impression, most memories are associated with a slew of other memories and general knowledge, all of which must lie in separate physical structures according to the physicalist model.

There is also the question of what stamps a memory as genuine rather than imagined. It cannot be the vividness or intensity of the experience, since hallucinations may be equally or even more intense than actual experiences. We seem to possess a higher mental function that can be called personal awareness or consciousness that allows us to place memories within the greater context of experience and decide whether they are real or imagined.

The renowned physicist and science philosopher, Henry Margenau (1901–1997), argued that the fact that different living entities all perceive the same world despite differences in their brains is evidence for the supremacy of consciousness. Since our only experience of the world is through our sense organs and brains, both of which differ greatly among individuals, it is remarkable that everyone perceives the same picture of the physical world. The differences are even greater in animals, yet they also seem to experience the same physical reality that we do. Margenau argued that this is only possible because we share the same consciousness or what he called the "One Mind." According to Margenau, this is the reason we perceive things the same way, and if this were not true then there would be many different perceptions of reality.[52]

Another problem is that of intention. Intentionality depends on an *I* or *user* as well as a symbol for what is intended. For example, we use language to express our intention to go to the store to buy a quart of milk. Inherent in our intentionality is a sense of *I am* and *I do*. The word milk in this case is a symbol for the type of object we intend to get. The user cannot be eliminated from the equation. The difficult problem is to come up with a theory of how intentionality can be produced by a physical process. Currently, no such theory has been proposed, and it has not been possible to produce intentionality by even the most advanced artificial intelligence.

Mind Effects on Body

Probably the most complete collection of evidence indicating that mind cannot be reduced to brain is contained in the book, *Irreducible Mind: Toward a Psychology for the 21st Century*.[53] The editor and one of the authors of this eight-hundred-page book include noted psychologist and professor in the Division of Perceptual Studies at the University of Virginia School of Medicine, Edward K. Kelly. Kelly and the other authors attempt to repudiate the conventional theory that the human mind and consciousness are material epiphenomena of brain. They point out that there are many physical changes that occur because of a change in a mental state that cannot be fully explained in terms of a physical brain process; and advance the idea—often associated with British psychologist Frederic W. H. Myers (1843-1901)—that mind must be a nonphysical entity independent of the brain and body.

The authors make a number of powerful arguments against neuroscientific reductionism, some of which are summarized below.

- The medical community now accepts the fact that mental states affect the body. Doctors are taught that symptoms of disease are sometimes only in the "mind" of the patient— psychosomatic illness. Psychological medical approaches, such as alternative medical treatments, a hands-on approach by a doctor or healer may be particularly effective in bringing about a cure of such patients.
- The placebo (and its counterpart the nocebo) effect is fully recognized as a way in which mental expectation can affect symptoms of pain and illness. The administration of a placebo, which by definition has no actual effect on the body, has a psychological effect on the individual. This in turn can cause a measureable physiological effect. The placebo effect is so well established that it is required to include a placebo in double blind clinical studies designed to show the efficacy of medical treatments.[54] Because the placebo effect is so potent, patients often show improvement from the placebo, and as a result, many experimental drugs and treatments have failed to show efficacy because of this.
- Psychological feelings of hopelessness and depression are strongly correlated with an increased risk of chronic diseases such as heart disease and cancer. On the flip side, experiences of joy and laughter have been demonstrated to improve health.
- Studies show that mental states can affect the immune system (psychoneuroimmunology). This is just one example of the close connection between mind and body. Today neuroscientists like to attribute effects like these to what they term: "mind-body unity." However, they lack a physical mechanism to explain such effects.
- It is known that there is increased mortality following bereavement. The stress and sorrow following the death of a loved one may cause a person to give up all hope and quickly die from cardiac arrest. In some cultures, the fear from receiving a curse

has similarly caused sudden death. On the other hand, there have been numerous reports of individuals that postponed their death until a significant event occurred in their life such as the birth of a grandchild. Research has shown that there is a positive correlation between positive emotions and health, and that such things as meditation, imagery, biofeedback, relaxation training, and hypnosis are effective for improving disease states.

- It is well known that hypnosis can produce surgical analgesia, changes in allergic reactions, and changes in autonomic functions such as heart rate, skin temperature, blood glucose, salivation, etc. Hypnotic suggestions can both cause and cure skin markings and blisters. Telling a hypnotized subject that they are holding their hand above a candle (when in fact they are not) can cause blisters to form on their palm. Conditions such as warts, eczema, psoriasis, and fish-skin disease (ichthyosis) have been cured by hypnosis.[55] Hypnosis as well as hysteria can cause temporary total blindness, deafness, and loss of speech.[56] It is difficult to explain how such specific physiological effects can be caused by the brain.
- Stigmata may be similar to hypnotic suggestion with the difference being that it is usually self-induced. This is where a person develops marks and sometimes bleeding wounds that Christ was thought to have suffered during his crucifixion. There have been hundreds of such cases reported in both the common and medical literature. One of the first such cases reported was that of St. Francis of Assisi. Persons with this affliction are most often intensely religious and have become emotionally tied to the suffering of Christ. Similar effects have also been observed in a nonreligious setting in which a person has bled from their hands, armpits, or eyes while undergoing extreme emotional stress.[57]
- A few exceptional individuals have been studied that can produce what is termed "skin-writing." One such person was Olga Kahl, who produced on her skin in less than a minute a communicated word or image. Her case was extensively studied and it was shown that the red color of the "writing" was well below the surface of the skin and required an exceptional control of peripheral circulation.[58] Neuroscientists are unable to offer a physiological explanation for how someone could have such exquisite control over capillary blood flow.
- Studies have shown that trained meditators and yogis can have extraordinary control over otherwise autonomic processes.

Examples include imperviousness to pain or cold, changes in skin temperature, and slowing or stoppage of heart and lungs. One such example taken from the medical literature was that of yogi Satyamurti who was fitted with a 12-lead electrocardiogram and demonstrated that he could stop his heart for seven days while practicing what he described as "nirvikalpa samadhi."[59]

- Multiple Personality or Dissociative identity disorder (DID) has been studied extensively by psychologists and is of interest because of the fact that the different personalities may exhibit different physiological conditions. For example, one personality may be allergic to a food and the other not. One may be right-handed and the other left-handed. One may have diabetes and require insulin and the other does not; or one requires glasses and the other does not.[60]
- There are numerous examples in which one person's mental state affected the body of another person. Examples include remote healing and "maternal impressions"—birthmarks or birth defects on a newborn that correspond to an unusual and intense experience of the mother during pregnancy.[61] There is also considerable evidence that prayer, with or without a person's knowledge, improves medical outcomes. Dr. Larry Dossey conducted ten years of research on the relationship between prayer and healing and found compelling evidence that it can complement medical treatments. The scientific evidence for this is outlined in his book, *Healing Words: The Power of Prayer and the Practice of Medicine*.[62]
- There are many cases reported of what is known as "terminal lucidity," in which mental clarity and memory return shortly before death in patients suffering from severe psychiatric and neurologic disorders. These include patients who are suffering from such serious brain disorders as Alzheimer's disease, stroke, brain tumors, and meningitis.[63]
- There are many examples of genius that arise from nowhere, persons with prodigious memory, calculating prodigies, and savants having spectacular abilities. To explain these phenomena requires one to hypothesize that their brains are "wired" differently from those of normal people. In addition, there are numerous cases of so-called "acquired savant syndrome" in which accidental or disease-caused brain damage led to newly acquired abilities, genius, and artistic skills.[64]

Many of the phenomena outlined above lack a physiological explanation and challenge the consensus paradigm of neuroscience that the brain *produces* consciousness. One would need to come up with a mechanism for translating an emotional sentiment into a specific neural pathway that would produce such physiological effects—but none is known. In addition, an explanation is needed for how damage to the brain causes new abilities to arise in an individual or for how lucid consciousness arises in brain-damaged individuals suddenly before death.

It is not enough to brush off such effects of the mind on the body by saying there is *mind-body unity*. One would have to explain the neuro-physical mechanism. How for example, could the expectation that one's hand is being burnt by a candle translate into neural pathways that damage skin cells and create a blister? We are forced to conclude that the physical explanation for mind is highly questionable because it struggles to explain many phenomena in which a mental state influences the body of a person—and sometimes that of another. In addition, it cannot even begin to explain qualia, self-hood, intention, and consciousness.

This data should increase our credence that materialism is a false ideology and that the spiritual view of reality might be true since all of these phenomena are consistent with the hypothesis that mind lies at a higher hierarchical level than matter and brain. In the next few chapters, we will explore other phenomena that directly contradict the materialist doctrine that mind can be reduced to brain. These include mystical experiences, out-of-body experiences, near-death experiences, psi phenomena (ESP), and memory of previous lives (reincarnation).

10

MYSTICAL EXPERIENCES

MYSTICS SOMETIMES DESCRIBE THE mystical experience as one of union with God. Alternatively, they describe the experience as a feeling of unity with the cosmos accompanied by a new understanding of their purpose in life. Although a salient feature of the mystical experience is its ineffability, the feeble attempts to describe it use words like clear, incredibly brilliant, timeless, unitary, ecstatic, and limitless consciousness.

Another common feature of the experience is knowledge that is nonlocal, intuitive, and not at all intellectual in nature. Mystics feel certain that they have been witness to a higher reality than that of everyday life. Such experiences are typically short-lived and may be spontaneous. They may also result from intense devotional sentiment, prayer, or meditation. The experience is normally transformative. That is, it has a profound and lasting effect. People report that the experience, no matter how transient, led to a profound change in how they view and conduct their lives—a change in attitudes, values, how they view death, and a new understanding of the meaning and purpose of life. And this change normally lasts for the rest of their life.

Psychologists recognize three states of consciousness: waking or normal consciousness, subconscious or dream state, and deep, dreamless sleep. Mystics claim that they have experienced a fourth state of consciousness in which there is the experience of unity with all things. This transcendental state of awareness is said to be as different from normal waking consciousness as the dream state. The fourth or mystical state of awareness is sometimes called cosmic consciousness.

What sets the mystical experience apart from other psychological states, such as hypnosis, hysteria, hallucinations, etc., is the feeling of unity with the cosmos (or God). One's consciousness is not tied to the body but seems to merge or become one with the whole of creation. Even individuals with a strongly religious background describe the experience as cosmic rather than religious. They might describe feeling that they were in the presence of their chosen savior, prophet, guru, etc., but in the full-blown mystical experience, qualifications based on religious preferences disappear into the *oneness of being*. These common features of the mystical experience are seen despite differences in age, culture, nationality, religion, and gender. This and the fact that the experience has a profound and life-changing effect provide good evidence that the experience is not illusory but a vision of a higher reality.

A Few Examples

One of the seminal texts on mystical experience was written by Canadian psychiatrist Richard M. Bucke in 1901, entitled *Cosmic Consciousness: A Revolution in the Study of the Human Mind*. Bucke himself had a brief experience at age 36 that led him to devote the rest of his life to investigating and reporting on the similar experiences of others. He relates his own experience that occurred in 1872 in his book as follows.

> I was in a state of quiet, almost passive enjoyment, not actually thinking, but letting ideas, images, and emotions flow of themselves through my mind. All at once, without warning of any kind, I found myself wrapped around as it were by a flame-colored cloud. For an instant I thought of fire, some sudden conflagration in the great city; the next, I knew that the light was within me. Directly afterward came upon me a sense of exultation, of immense joyousness accompanied by an intellectual illumination quite impossible to describe. Among other things, I did not merely come to believe, but I saw that the universe is not composed of dead matter, but is, on the contrary, a living Presence. I became conscious in myself of eternal life...I saw that all men are immortal; that the cosmic order is such that without any peradventure all things work together for the world, is what we call love, and that happiness of each and all is in the long run absolutely certain.[65]

He went on to write that, "I learned more within the few seconds that illumination lasted than in all my previous years of study and I learned much that no study could ever have taught."

Another example of a mystical experience was reported by Robert Adams (1928-1997) who was born and educated in the United States. At the age of fourteen, he had his first mystical experience, which forever changed his life, and by the age of sixteen, he began an earnest quest to find a spiritual teacher.

> When I had my spiritual awakening, I was fourteen years old. This body was sitting in a classroom taking a math test. And all of a sudden, I felt myself expanding. I never left my body, which proves that the body never existed to begin with. I felt the body expanding, and a brilliant light began to come out of my heart. I happened to see this light in all directions. I had peripheral vision, and this light was really my Self. It was not my body and became brighter and brighter and brighter, the light of a thousand suns. I thought I would be burnt to a crisp, but alas, I wasn't. But, this brilliant light, which I was the center and also the circumference, expanded throughout the universe, and I was able to feel the planets, the stars, the galaxies, as myself. And, this light shone so bright, yet it was beautiful, it was bliss, it was ineffable, indescribable. After a while, the light began to fade away and there was no darkness. There was just a place between light and darkness, the place beyond the light. You can call it the void, but it wasn't just a void. It was this pure awareness I always talk about. I was aware that I AM THAT I AM. I was aware of the whole universe at the same time. There was no time, there was no space, there was just I am.[66]

Naturally, religious overtones play strongly in the numerous worldwide accounts of these experiences. In fact, most of the world's great religions grew from the mystical visions of such people as Shiva, Lao Tzu, Moses, Buddha, Jesus, Mohammed, and Baha'u'llah. Although the original mystical message of such prophets may have been diluted or obscured by centuries of religious doctrines and rituals, certain core mystical traditions have survived. These include the yogic and tantric practices of Vedantism, Buddhist meditative practices, Kabbalistic Judaism, Christian mysticism, and Islamic Sufism.

Apparently, religious conviction is not a prerequisite for having a mystical experience. The experience can arrive spontaneously in people that have no strong religious beliefs or inclinations. For example, William James in his important study of mystical experiences entitled, *The Varieties of Religious Experience*, describes the experiences of a number of secular-minded individuals including Lord Tennyson, who described his experience while in solitude:

> ...all at once, as it were out of the intensity of the consciousness of individuality, individuality itself seemed to dissolve away into boundless being, and this not a confused state but the clearest, the surest of the surest, utterly beyond words—here to death was an almost laughable impossibility—the loss of personality (if so it were) seeming no extinction, but the only true life. I am ashamed of my feeble description. Have I not said the state is utterly beyond words?[67]

Drug-Induced Mysticism

There is no question that psychedelic drugs can produce many of the same experiences described by mystics. There is a long list of studies linking the use of LSD, mescaline, and psilocybin to the full range of mystical experiences.[68] This has led many neuroscientists to label all mystical experiences as a product of abnormal brain chemistry. Such an analysis might be correct, but it would not explain the nature of the experiences and why they are so transformative. Whether the experience is brought on by some outside chemical agent or endogenous alteration in brain chemistry is irrelevant. The experience is clearly a *different conscious experience*. An analogy might be the difference between a pond (normal waking consciousness) and an ocean (unlimited consciousness).

The most popular theory of how psychedelics exert their effect is the "filtering model." One of the first to propose this theory was English author and philosopher, Aldous Huxley, who described his own psychedelic experiences with mescaline in his book, *The Doors of Perception*.[69] According to this model, the brain normally filters out most of the sensory and cognitive input that is affecting it at any given time. When this "filter" is largely removed by the psychoactive drug, the mind becomes overwhelmed and experiences a unique transcendental state of consciousness. We connect with a new vision of reality that is always there, but normally off-limits to the mind as it engages in the information processing of normal life. The model suggests that there is, as Huxley called it, "Mind at Large." Other names used to describe this universal mind are one mind and cosmic mind. In any case, in order to make biological survival possible the brain attenuates our awareness of this vast reservoir of consciousness to a mere trickle, without which we would be unable to function. Psychedelic drugs (and presumably mystical states) open this valve allowing us to experience a much-enhanced version of awareness.[70]

This model for how the brain normally attenuates conscious awareness is further bolstered by the fact that fMRI scans of people experiencing altered and highly enhanced states of consciousness following the administration of psilocybin, actually showed *decreased* blood flow in specific regions of their brain corresponding to decrease neural activity, but interestingly, *no regions were observed to have increased flow.*[71]

My first mystical experience occurred when I was 22 years old and was aided by the use of methylenedioxymethamphetamine (MDMA). This is how I would describe it:

> About three months earlier, I began practicing yoga and meditation. One evening I took the drug, lit a candle and some incense, and put on some music. Then I sat comfortably in half-lotus posture and began my meditation. After about twenty minutes, I began to feel a wonderful sensation at the base of my spine that grew stronger and stronger and pulsated up my spine bringing me greater and greater bliss. As this extremely pleasant sensation grew in intensity, I perceived the energy originating at the base of my spine gushing upward, flooding my brain with ecstatic warmth and light. These sensations were accompanied by a deep roar like that of the ocean, and then suddenly with a flash—like lightning, a huge burst of energy came straight up my spine and lit up my brain in indescribable light. Every vibration I had been experiencing seemed to cease and I felt my being become nothing but infinite peace-space-bliss. It felt as though the entire universe was within my being in a state of perfect peace and balance. The only thought in my mind was "I exist and am the One." I do not know how long I remained in this state of union, but the next

morning I felt myself coming out of the trance and began to cry, having lost this state of blessedness. The next day or two it was impossible for me to work or do much of anything that required my attention to the outside world.

The experience changed my life forever. From then on, I knew that getting back to that state was the only meaningful goal in life. I began an earnest search for a spiritual teacher that could instruct me on how to obtain this experience without psychedelic drugs.

A Timeless Experience

In religious scripture, the most concise description of the mystical state might be that of the Mandukya Upanishad. It says that below the waking state lie the states of dreaming sleep and deep dreamless sleep. But beyond these is the fourth state (*turiya*), the transcendental state, which is characterized by pure unitary consciousness and bliss (*ananda*)—invisible, otherworldly, incomprehensible, without qualities, indescribable, the unified soul in essence, peaceful, auspicious, and without duality.

Mystics often describe their experience as entering into a "timeless state" or "eternal now." In other words, they describe the stoppage of the flow of time. Similar to my own experience, it is common for mystics to describe feeling that the entire universe lies within their being; and there is often the realization that the cosmos is the thought projection of the Godhead. For what may last only a moment, time stops and the boundaries of their being dissolve into the One—there is no longer any difference between them and God. Universally they describe coming to the realization that our perception of reality is illusory—a product of perceiving a relative reality. Ultimate Reality is both limitless and timeless.

Interestingly, the mystic's perception of reality—which can go by the name of cosmic consciousness—is exactly what we might expect if a person was to experience the entirety of spacetime as described by Einstein's theory of relativity. Witnessing the full four-dimensional continuum of spacetime, we would no longer expect them to perceive the illusory flow of time. All time would be there, unchanging within the conscious awareness of such a vision of reality.

Summary

There have been hundreds of accounts of mystical union described in the popular literature. The experiencers universally say they entered into an indescribable state of conscious awareness that was unitary, ecstatic, and timeless. In this state, they report that their little self (ego) disappeared as they merged or became one with cosmic consciousness. Their descriptions of this life-changing event are strikingly similar. Attempts to label their experience as merely hallucinations or products of abnormal brain chemistry would have to come to terms with this fact, and the fact that the experiences are both incredibly powerful and transformative. Mystics claim to have experienced a state of consciousness that is more expansive, knowing, and unitary than normal waking consciousness, and the experience is so profound that they feel certain that consciousness is the foundation from which reality is built. In other words, they see creation as an ongoing unfolding of consciousness into the material world.

The mystic's vision of a transcendental reality is contrary to the physicalists' doctrine that mind and consciousness are byproducts of matter; but unfortunately, the experiences of mystics are personal and cannot be shared directly with us so there will always be skeptics who will label such experiences as "anecdotal" or products of a deranged or abnormal mind. However, the mystical experience is compatible with the modern scientific view of the cosmos. It only requires that we imagine breaking out of the confinement of our three-dimensional way of perceiving reality and entering a state in which we perceive wholeness—all four dimensions of spacetime simultaneously. No matter how we might weigh the evidence of whether such personal experiences indicate the existence of a "higher" level of consciousness, it should increase our confidence that the spiritual worldview is a better model for reality than the physicalist version.

11

OUT-OF-BODY AND NEAR-DEATH EXPERIENCES

Can Consciousness Leave the Body?

DURING NORMAL OR WAKING consciousness, we have a clear sense that we are *in* our body. This awareness is altered in what is called an out-of-body experience (OBE). Here a person has a vivid experience of *leaving* their physical body, yet they experience normal and sometimes more lucid consciousness. They describe the sensation of floating above their body and are able to witness events from this unusual perspective. The OBE is a common feature of the near-death experience (NDE), but differs in that the OBE is usually not associated with a close brush with death. OBEs are most likely to occur during anesthesia, while falling asleep, and during lucid dreams. Psychologists label the OBE a dissociative experience arising from abnormal psychological and neurological factors—an altered state of consciousness like a dream or hallucination.

Few neuroscientists today consider the OBE to be evidence for the ability of the mind to leave the body and gain information that would not be available locally through the sense organs. Investigators also report that many of the features of an OBE can be reproduced using artificial stimulation of the brain and by psychedelic drugs such as LSD. The consensus among neuroscientists is that OBEs are of physiological origin. But is this correct? While some of these experiences may have a psychophysiological explanation, there are good reasons to believe that a paranormal explanation is needed for some OBEs.

One reason is the accuracy in which people with OBEs are able to describe attempts to revive them when their heart was stopped, and according to the physicalist model, they should have been unconscious—certainly incapable of having lucid consciousness. There are also cases of persons with total blindness having visual experiences that have time stamps indicating that they witnessed events when there was cessation of blood flow to the brain.

An example is a woman—blind from birth—who is admitted to the hospital suffering from acute appendicitis. While under anesthesia to remove her appendix, she suddenly develops ventricular fibrillation. The hospital staff rush to her bedside to try to restore her normal heartbeat since this condition halts blood circulation to her brain and she will certainly die. The hospital staff uses a defibrillator to restore her heart to normal sinus rhythm, and completes the operation without further complications. Following the procedure, she tells the medical staff and members of her family that she experienced the sensation of floating above her body, and from that vantage she could clearly hear and see the things that occurred as the doctors and nurses frantically tried to restart her heart. Despite being blind and anesthetized, she could accurately describe details of the operating room and the efforts that were made to resuscitate her.

There have been literally thousands of people who have reported observing events from a perspective outside their body. H. Hart did an analysis of almost three hundred published accounts of OBEs in which the person reported observing events that they could not have witnessed using their ordinary senses, and the details in nearly one hundred of these cases were corroborated by a second party or by other means.[72] Hart also investigated OBEs in which a second person at a distant location had a vision or felt the presence of the person undergoing the OBE. Such cases are called "reciprocal apparitions." If the knowledge received by persons undergoing an OBE were strictly due to an ESP phenomenon known as remote viewing (clairvoyance), then it would be difficult to explain how another person could share in the experience.

Some people report that while having an OBE they became aware of events that occurred at a distance outside the range of normal senses, and these events were verified by others.[73,74] Such reports are particularly difficult to explain using the physicalist model that label these experiences as psychophysical abnormalities.

Dr. Sam Parnia is a leading expert on the scientific study of death. Beginning in 2001, he and colleagues have investigated claims of OBEs by persons undergoing cardiac arrest. These so-called AWARE studies (AWAreness during REsuscitation) have looked into the experiences of over 1500 cardiac-arrest survivors in order to determine whether people without a heartbeat or brain activity can have documentable OBEs and near-death experiences. About 2 percent of the individuals in the study reported full awareness compatible with an OBE. Most of the patients that experienced an OBE described having clear visual experiences from a vantage point outside their body—something Parnia claims would be impossible when the brain is shut down during cardiac arrest.[75] In fact, he claims most patients who have had cardiac arrest report experiencing no recollection of the incident and such amnesia is consistent with what we know about an oxygen-starved brain.

For adepts in yoga, the out-of-body experience is taken for granted as one of the psychic powers (*siddhis*) that may be acquired through intensive meditation. The term used to describe this supernormal power is "astral projection." It is believed that the psychic body leaves the physical body but is still tied to the physical body by a subtle cord—much like a kite is attached to the ground by a string. The psychic body is free to roam the cosmos according to the will of the yogi.[76] Similar to other powers obtained by spiritual practices, students of yoga are constantly reminded not to pursue them because they can lead one astray and be dangerous to their psycho-spiritual health.

Near-death Experiences

The near-death experience is characterized by a lucid out-of-body experience along with feelings of peace and bliss. Similar to the OBE, there will usually be accurate visual and auditory experiences from a vantage point outside the body. Often people report seeing their lifeless body from above and witnessing attempts to resuscitate it. Next, there is usually the experience of entering a dark tunnel with a brilliant light beyond. Sometimes loved ones are seen in the tunnel. Once they enter the light at the end of the tunnel, there is the feeling of being in the presence of a being that radiates infinite, eternal, and unconditional love.

NDEs are normally associated with a close brush with death, such as cardiac arrest, but may also be triggered by a strong expectation that they are going to die but in which there is actually no possibility of death. During the NDE, they claim to experience being in a nonphysical body that is completely healthy, pain free, weightless, and blissful. They report being fully conscious and have full memory, judgment, and imagination. The images they witness in this "disembodied state" are described as highly vivid, with lucid awareness and clarity—often described as more real than normal waking consciousness.

Often these experiences are complimented by a life review in which the dying person experiences an altered state of consciousness in which they have an omni-view of life events. It has been described as witnessing thousands of events in their life simultaneously, each consisting of a separate scene from their life—nothing like a dream or sped-up movie. Often the events are not what they might have considered as the more important ones in their life, but trivial events where they see how their actions affected other people both positively and negatively. NDErs say they no longer have any fear of death, but many say they realized that their work on Earth is not complete and that they may need to return to complete their journey to attain final union with God.

Finally, the person may feel that there is a line that if crossed will lead to death. They either feel or are told that it is not their time to go and describe being reluctantly drawn back to their body—the usual reason being commitment to family or loved ones. Almost universally, the person will describe the NDE as a life-changing event. Their attitudes, beliefs, and outlook on life and death are dramatically and permanently changed. Even among persons who were previously atheists, there is a certainty that God exists and that there is life after death.

The history of NDEs goes back thousands of years and they have been reported in many diverse cultures. One of the earliest accounts was mentioned by Plato in *The Republic*. He tells the story of a soldier named Er who was fatally wounded in battle and being thought dead was placed on a funeral pyre, only to awaken and describe in detail his experience. Er described leaving his body and traveling to what he described as the afterlife. There he witnessed his past deeds and felt that the final destination of the soul was determined based on those deeds. He met with loved ones and friends long dead and experienced the great beauty and brilliance of a place "filled with great radiant light." Then he described being drawn back to his body and awakening in pain but having the knowledge that death was not final, but merely a transition.

Paul's experience on the road to Damascus sounds like it could have been a NDE. In modern times, numerous such reports have surfaced both in the popular and medical literature. Estimates place the prevalence of NDEs at 10–20 percent of patients close to death,[77] while the Near-Death Research Association (NDFR) estimates that 5 percent of the American population has experienced an NDE.[78]

Raymond Moody has probably done more to popularize the NDE than any other researcher. In his bestselling book, *Life after Life*, he chronicled the accounts of 150 survivors of a near-death experience and first coined the term near-death experience to describe these accounts. Moody concluded that the NDE was strong evidence for the continuation of consciousness following death of the body.

Thousands of people have had near-death experiences, but scientists have argued that they should not be considered as paranormal. Dr. Eben Alexander was one of those scientists. An academic neurosurgeon for 25 years at Harvard Medical, Alexander knew that NDEs feel real, but are fantasies produced by brains under extreme stress. That was until Dr. Alexander's own brain was attacked by a rare form of meningitis caused by the *E. coli* bacteria. This caused the part of his brain that controls thought and emotion—and in essence makes us human—to shut down completely. For seven days, he lay in a coma. Then, as his doctors considered stopping treatment and removed his ventilator, Alexander's eyes popped open. He had come back.

Alexander's recovery was a medical miracle. But the real miracle of his story lies elsewhere. While his body lay in coma, Alexander had a near-death experience that convinced him that his physicalist approach to mind had to be wrong. During his NDE, he encountered a "Divine Presence" that was the source of the universe itself—far beyond any human consciousness.

Alexander's story is unique because it happened to a scientist who has a deep understanding of the brain and how it works. Before he underwent his NDE, he could not reconcile his knowledge of neuroscience with any belief in heaven, God, or the soul. Today Alexander is a doctor who is convinced that there is definitely something about consciousness that our primitive models don't get. God and the soul are real and death is not the end of our personal existence—only a transition. He relates his amazing journey into the afterlife in his bestselling book, *Proof of Heaven: A Neurosurgeon's Journey into the Afterlife*.[79]

Another researcher into NDEs is Jeffrey Long. In his book *Evidence of the Afterlife: The Science of Near-Death Experiences*, Long argues that there are nine observations that prove the existence of life after death.[80] These were generated through his study of hundreds of NDEs and the consistencies of the accounts that he compiled over the years. His arguments include that it cannot be medically explained how people experience consciousness outside their body when they are clinically dead; blind people experiencing visual perceptions during their NDE; children giving details of their NDEs similar to adults, though they may have never been exposed to this concept; and "life-review" experiences that reflect real events. These observations are the primary basis for Long's assertion that the NDE data proves that there is life after death.

One of the most compelling reasons for believing that the NDE is a real phenomenon associated with the dying process is the remarkable similarity of the accounts, regardless of age, nationality, religion, race, culture, and other demographics. No two experiences appear identical, but even among children, one or more of the elements mentioned above are reported.

However, if we accept the reality of the experience, does it really negate the prevailing opinion of neuroscientists that all mental activity can be attributed to the brain? I believe the evidence is overwhelming that it does. What is most difficult to explain is how there can be a continuation of consciousness, and even an enhancement of mental awareness during a time when the brain is shutting down or which in many cases has ceased functioning completely because of a stoppage of blood flow. Not only are there lucid consciousness and vivid memory, but also fully structured thought processes and the same sense of self that exists in normal waking consciousness. Often the experience is described as so incredibly beautiful and transcendent that words simply cannot describe it. In addition, the effect on the person may be felt for decades.

EEG studies of people suffering cardiac arrest indicate the absence of gamma waves, normally associated with waking consciousness, and within a few seconds of circulatory collapse the EEG will display a flat line, which is one of the characteristics of death—the others being lack of heart beat, respiration, and brainstem reflexes. Cardiologists and neuroscientists agree that the oxygen-deprived brain of persons suffering cardiac arrest could not produce the images and lucid consciousness that is a hallmark of the NDE. In fact, the ordinary unconsciousness that accompanies such events is typically associated with confusion and impairment of memory (amnesia).

It is also very difficult to explain how persons undergoing such insults to the brain could obtain and later relate accurate and verifiable information about events that took place while they were unconscious and in some cases considered clinically dead. This does not mean that neuroscientists have not tried to offer a physicalist explanation for the NDE.

One theory is that during the process of dying, as the brain shuts down, people may experience a dream-like state and/or hallucinations as the subconscious mind takes over. However, this theory fails to adequately explain how people who are pronounced clinically dead could have such lucid consciousness, a consciousness that is described as more vivid than normal day-to-day consciousness—nor does it explain many of the other elements of the full-blown NDE. In fact, hallucinations are usually illogical, fleeting, bizarre, and provide a distorted sense of reality, whereas the vast majority of NDEs are logical, orderly, clear, and comprehensible. Like dreams, people tend to forget their hallucinations, whereas most NDEs remain vivid for decades. Furthermore, NDEs often lead to profound and permanent transformations in personality, attitudes, beliefs, and values—something that is never seen following hallucinations. People looking back on hallucinations typically recognize them as unreal, as fantasies, whereas people often describe their NDEs as "more real than real." Further, people who have experienced both hallucinations and an NDE describe them as being quite different.[81]

Some skeptics have postulated that a flat-line EEG does not guarantee lack of brain function. There is ample evidence to suggest that there can be deep-seated neuronal activity in the brain that does not show up on a scalp EEG. Hence undetected brain activity could be going on and be responsible for the NDE. The problem with this theory is not that there might be brain activity, but whether there could be the type of brain activity that is considered necessary for conscious experience. The answer appears to be no, since the hypothesis that there is a brain basis for consciousness depends on the ability of widely separate regions of the brain to coordinate with one another.

It has been suggested that some of the features of a NDE can be reproduced by drugs or other means. For example, the massive release of endorphins (endogenous opioids) in the brain might be responsible for causing blissful feelings and an altered state of consciousness. And drugs like ketamine, DMT, psilocybin, and LSD are known to cause altered states of consciousness. However, according to Dr. Alexander, the experience is totally unlike any drug experience. He describes such drug experiences "as not even in the right ballpark."

One of the biggest challenges facing such skeptics is how any of the brain-based or physiological explanations for the NDE can produce experiences that seem totally real, memorable, provide accurate images of events that took place while a person was unconscious, include a life review, and have lasting and profound aftereffects. None of the proposed physiological explanations offered by skeptics comes close to producing even one of these effects, let alone the full spectrum of them. In fact, the one common denominator in most altered states of consciousness having mystical, out-of-body, and transpersonal qualities is that they occur when there is a *decrease* in brain activity due to a decrease or stoppage of cerebral blood flow, which would be consistent with the filtering model for how the brain normally restrains conscious awareness.

It is also hard to explain how someone that is not in any danger of dying could have similar experiences. The term for this is near-death-like experiences (NDLE). One such experience is laid out in detail by Nancy Clark in her book *Divine Moments*.[82] In the early 1960s, she had a near-death experience in connection with the birth of her son, where she was actually left for dead. She experienced all of the classical elements of the NDE: leaving her body and seeing medical personnel trying to revive her, light streaming toward her, peace, bliss, universal and unconditional love permeating her being, and a life review. This was well before people started talking about NDEs and she did not tell a soul about her experience, fearing people would think she was crazy. Then 17 years later, she had an almost identical experience while standing at a podium in front of a group of people delivering a eulogy for a dear friend. She described this experience as being "merged into Oneness with the Light of God." After about 15 minutes in this ecstatic state of oneness, she reluctantly returned to her body while not missing a beat in her eulogy.

After her NDLE, Clark has spent the last thirty years researching spiritually transformative experiences that happen to ordinary people—most of whom, like her, were in no danger of dying. She points out that the term near-death experience can be misleading since perfectly healthy people can have a NDE-like experience—just like hers. And she is in a good position to know this because she underwent both types.

A further challenge to the brain-based explanation of the NDE comes from Raymond Moody. He points out that elements that we think of as a near-death experience—leaving one's body; seeing deceased relatives; going into the light and feeling unconditional love; and seeing a panoramic review of one's life—occur not just to people who have the NDE and are brought back—but to healthy and uninjured bystanders at the bedside of the victim. These so-called "shared death experiences" are not infrequent. The bystanders say that they also left their bodies and accompanied their dying loved ones partway toward the light, or that they saw the apparitions of relatives and friends of the dying person. Moody has accounts of hundreds of people with shared death experiences—some of whom had the identical experience to the person with a NDE. They reported experiences characteristic of the NDE but they themselves were perfectly healthy.[83] If the cause of these near-death experiences were a physiological insult to the brain, then how could bystanders who were not ill or injured have essentially the same experience?

Summary

OBEs and NDEs belong to a larger family of mystical or transpersonal experiences that transcend the usual limits of space and time and may potentially be spiritually transformative. They can occur spontaneously or be triggered by a number of factors including, for the NDE, a close brush with death—but this is not a requirement.

The popular current hypothesis of neuroscience is called epiphenomenalism, which in effect says that consciousness is dependent on the electrochemical activities of the brain. This is despite the fact that neuroscience has yet to propose a meaningful model for how consciousness and self-awareness arise from strictly neurochemical processes in the brain. Most neuroscientists freely dismiss the OBE and NDE as aberrant brain conditions, and dismiss all the evidence to the contrary as anecdotal, superstition, bad science—or simply impossible. The history of science has many examples in which the vast majority of scientists favored the status quo even in the face of anomalous evidence that in hindsight should have offered clear indications that the prevailing theory was incorrect.

However, there is a huge volume of data on OBEs and NDEs that indicate that these experiences involve altered states of awareness with vivid and unforgettable images and sensations—mental functions that could not possibly occur according to current neurophysiologic models of how the brain produces mind. The alternative hypothesis that mind is nonmaterial and may separate from the body under certain conditions, including death, appears to be a more plausible explanation for these experiences.

12

PSI PHENOMENA

THE TERM PSI PHENOMENA refers to extrasensory perception (ESP), but researchers in this field, known as parapsychology, prefer the moniker psi to describe these phenomena because they feel the term ESP implies that it is an "extra" sense when in fact almost all people can be shown to have this ability.

Psi phenomena are normally grouped into four related phenomena: clairvoyance (or remote viewing), telepathy, precognition, and psychokinesis. Clairvoyance is the ability to obtain information about places, things, or events at a remote location. Telepathy is the transfer of thoughts or feelings between individuals at a distance without any apparent physical means. Precognition involves perceiving information about future events before they occur, and psychokinesis is the ability of the mind to influence matter, spacetime, or energy.

There are words to describe the various psi phenomena in every major language of the world suggesting that psi is a universal phenomenon, and that it is a basic element of human experience. However, serious scientific research into psychic phenomena has been going on for only about 140 years. Carefully controlled studies of telepathy, remote viewing, precognition, and psychokinesis began in the 1930s when researchers at Duke University became the first major U.S. academic institution to engage in the critical study of psi in a controlled laboratory setting. Because of the psi experiments at Duke, standard laboratory procedures for the testing of psi developed and came to be adopted by interested researchers throughout the world.

A number of leading parapsychology researchers have written books detailing the thousands of well-controlled and reproducible results that they claim prove beyond any reasonable doubt that psi phenomena are real.[84,85,86] I will not attempt here to present anything more than a brief summary of some of the data that gives researchers in this field the confidence to state that the scientific method has "proven" the psi is real. Of course, what they mean is that if anyone were to consider the overwhelming evidence for these phenomena, it would be logically unreasonable to deny the reality of it. Studies on psi have been going on for more than a century, and even a die-hard skeptic would have to admit that if there was nothing there, researchers in the field would have discontinued such studies long ago. With these things in mind, let's dive into some of the evidence beginning with that for telepathy.

Telepathy

The word telepathy actually means "feeling at a distance." However, since it was first coined by renowned psychologist F. W. H. Myers in 1882, it has come to mean communication between two minds.

The first well-controlled series of telepathy experiments were conducted by Joseph Rhine and his colleagues at Duke University from 1927 to 1965. Rhine developed a method using a special deck of twenty-five cards with five different symbols (square, circle, wavy lines, star, and triangle). A sender shuffled the cards and starting from the top of the deck attempted to "send" the symbol of the card to a remote person using a prearranged timing scheme. The "receiver" was required to make a selection and by chance would be expected on average to get five cards out of twenty-five correct. When Rhine published the results of 188 experiments in his 1940 book, *Extrasensory Perception after Sixty Years*,[87] he showed that the odds that the results were by chance were less than one in a trillion. In a psi card test of this type, it is impossible to know if the receiver was getting information via telepathy or clairvoyance.

Today the research method of choice for the demonstration of telepathy is the ganzfeld method, which means "complete field" in German. It is well established that a quiet, peaceful environment with limited sensory inputs is optimal for receiving information from another mind. Hence, the receiver in such experiments is situated in a comfortable reclining chair with halved ping-pong balls placed over their eyes and white noise playing on headphones. A soft red light illuminates their face. These conditions create a restful state similar to one produced in a sensory deprivation chamber or floatation tank. Next, an assistant is asked to select one target pack of four images from a pool of many such packets. Then one target image is selected from the four images in this packet, which is in an opaque envelope. This envelope is given to the sender, who is in a distant, isolated room called the "sending chamber." The sender unseals the envelope and mentally tries to send the image to the receiver over the next fifteen to twenty minutes, after which the receiver is returned to a normal environment and asked to select the image that was "sent" from the four images that were in the original packet. Variations to this procedure include using video clips and automating the selection process using a computer. The receiver is asked to rank the images from one to four. If the actual image (or video clip) is ranked first, it is considered a hit, which is expected to occur by chance one fourth of the time.

From 1974 to 1997, there were over 2500 ganzfeld sessions reported in forty publications by researchers around the world. In a 1985 meta-analysis[88] of the published studies that provided hit data, twenty-three out of twenty-eight studies resulted in hit rates greater than chance with odds against chance of a billion to one.[89] After receiving some criticism from skeptics, a new round of experiments was started in 1983 by Honorton and colleagues that were computer-controlled. During the six years of the study using 354 volunteers, a 34 percent hit rate was achieved compared to the 25 percent expected by chance.[90] These results were similar to the nonautomated experiments indicating that people can sometimes receive information at a distance without the use of the five senses.

There have been numerous, well-documented reports of telepathy between identical twins—some separated by birth and later reunited. In one famous case sited by Russell Targ, two identical twins were separated after birth and both named Jim. Although they did not communicate until thirty-nine years later when they met for the first time, they both married a woman name Betty, divorced her, and then married a woman named Linda. Both were firefighters, and coincidently both had built a circular white bench around a tree in their backyard.[91] In addition, there are many stories of twins experiencing unusual but specific impressions, feelings of apprehension, pain, and concern for their twin brother or sister that later turn out to reflect actual events that occurred around the same time.[92] The only reasonable explanation for the uncanny connection that clearly exists between some twins is that their minds are entangled, just as their bodies were in the womb.

Remote Viewing

The difference between remote viewing (clairvoyance) and telepathy is that no sender is required. Information, normally in the form of images, is obtained about something from a distance.

The five-symbol card decks used by Rhine and his colleagues at Duke University also became a standard for investigating remote viewing. The cards were in sealed envelopes, behind opaque screens, separated by distance, or even displaced in time. The hit rate using all these methods in a subset of tightly controlled card experiments run by twenty-four different investigators from 1934 to 1939 in over nine hundred thousand trials was significantly above the 20 percent expected by chance.[93] J. G. Pratt calculated that if all the clairvoyance card tests conducted from 1882 to 1939 were combined, the odds against chance would be more than a billion trillion to one. This was a four-million-trial database performed by sixty investigators in many different countries and reported in nearly two hundred published reports.[94]

A program begun in 1972, during the height of the cold war, was conducted at the Stanford Research Institute (SRI) to determine whether remote viewing could be used for espionage purposes. This was a twenty-million dollar, twenty-three year program funded by the CIA, NASA, and agencies of the Department of Defense. In these experiments, a talented or gifted viewer was asked to sketch or describe a target facility. In most studies, a sender visited the site and attempted to send mental images of the site to the viewer telepathically. For testing purposes, a secret facility in the United States was sometimes used, and for espionage purposes, the viewer was asked to describe a hidden facility in a foreign country in order to enhance and corroborate other intelligence about it. In 1976, SRI researchers Harold Puthoff and Russell Targ published their initial results using primarily one highly gifted viewer—Ingo Swann. Swann would describe his mental impressions of the site visited by the "sender" and his description would be carefully judged against the description of 6-10 randomly chosen additional locations in a double-blind manner. The results of these trials were highly statistically significant with odds of a chance occurrence of one in a hundred thousand.[95]

In the years that followed SRI, researchers Puthoff and Targ identified several other gifted "viewers," and by 1989, the program was moved to the

Science Applications International Corp. (SAIC)—a defense contractor, which continued the research until 1994.

In 1988 May and colleagues analyzed all the psi experiments conducted at SRI over a sixteen-year period, most of which were remote viewing tests. The data set consisted of over twenty-six thousand trials. Their statistical analysis indicated odds against chance of 10^{20} to one.[96]

The SRI results were successfully reproduced by Princeton University Professor Robert Jahn and Brenda Dunn. Students in the laboratory were asked to describe their mental impressions of a site that was randomly selected at which a person was hiding. Their findings, which comprised over 400 trials and several years, showed that increased distance between the viewer and sender had no effect on the viewer's ability to describe the site, nor did it matter if the sender was at that moment at the site or had visited it a week earlier. Their final report published in 1982 showed that the odds that chance would explain the results were one in a billion.[97]

In his book, *The Reality of ESP*, physicist Russell Targ describes the more than 40 years of published experimental evidence that convinces him that ESP is now scientifically proven, and a reasonable person should no longer doubt its reality.[98] An example of some of the evidence is summarized below.

- A remote viewer with Army Intelligence, Joe McMoneagle, located a downed Soviet bomber in Africa and described the secret construction of a 500-foot Soviet Typhoon-class submarine in Russia.[99]
- Among his many accurate descriptions of distant locations, Ingo Swann, in 1973 described a ring around the planet Jupiter in a remote viewing session. At the time, Jupiter was not known to have any rings, but the ring was later confirmed by the Voyager probe in 1979.[100]
- In 1974 retired police commissioner, Pat Price, described and drew to scale a Soviet weapons factory in Siberia with remarkable detail. What he drew was confirmed by satellite photography.[101] Price also identified the kidnapper of Patricia Hearst from a mug book containing hundreds of photos, and told police where to find the kidnapper's car.[102]
- In 1982, a small group of investors used remote viewing to forecast successfully changes in the price of silver nine weeks in a row and made $120,000. According to a knowledgeable commodity trader doing this in a volatile market is next to impossible.[103]

Parapsychology researcher, Stephan Schwartz, has authored several books and research papers on the use of remote viewing in archaeology. In 1977, he elicited the help of Ingo Swan along with another psychic, Hella Hammid, who using remote viewing both independently identified within a few hundred yards on a chart the location of a century-old wreck lying hundreds of feet below the surface near Santa Catalina Island off the coast of southern California.

Details of this so-called Project Deep Quest are provided in his book, *Opening to the Infinite: The Art and Science of Nonlocal Awareness*.[104]

The Princeton Engineering Anomalies Research Laboratory (PEAR) conducted over 650 remote-viewing trials. The protocol required one participant to be stationed at a randomly selected location at a given time and to observe and record impressions of the details and ambiance of the scene. A second participant, the "viewer," located far from the scene and with no prior information about it, tried to sense its configuration and character and to report these in a similar format to the sender's description. Using a numerical-scoring method to evaluate the accuracy of viewer's subjective descriptions and the physical targets, they concluded this would occur by chance only three times in ten billion.[105] The data confirmed the hypothesis that there exists an unexplained mode of information acquisition that is independent of both time and distance between the viewer and the target.

Precognition

Precognition, as the name implies, means knowing something before it happens. Precognition might also be called fortune telling or prophesy. History is marked by innumerable examples of people who have accurately prophesized their own imminent death, a disaster, or other future events. Various cultures and religions take prophesy quite seriously, and it forms the basis for many mythologies. Even today, while it is common to dismiss many of the stories of lore as myths and superstitions, it is such a common experience for people to have an intuitive feeling about something that later happens, that prophesy and fortune telling are accepted by a majority of people worldwide. This does not mean precognition is true, but it does suggest that it may not be that rare for a person's mind to bump into their future.

However, the experimental "proof" of precognition comes from thousands of well-controlled scientific experiments. The typical experimental protocol is to have a person guess in advance which target will be displayed from a randomly selected group of possible targets. If the correct target is selected, then this is counted as a hit. Many studies of this type have been conducted since 1935, and in almost all of the studies, the null hypothesis (results are due to chance) was rejected. Combining the results from 309 studies, 113 articles, sixty-two different investigators, and almost two million individual trials gave odds against chance of 10^{25} to one.[106] Odds of even ten thousand to one indicate that chance occurrence must be eliminated as an explanation for the results.

Experiments also show that unconscious autonomic responses to stress occur before the stressful situation is witnessed consciously. The technical term for such a response is "presentiment." It can be measured by attaching sensors such as skin conductance, heart rate, or blood flow in the fingertip, and then having a computer display a series of photos, some of which are calming and some of which are disturbing. Naturally, the monitors show a strong response immediately after the subject sees a disturbing photo unlike that following a calm photo—such as a field of flowers. The surprising thing is that the computer registers a significant fright-and-flight response just before the emotionally charged photo is shown. The opposite effect of a dropping heart rate can also be seen just before the calm photos are shown.[107] Experiments like this in which an unconscious autonomic response precedes the actual insult have been replicated and indicate that the nervous system is capable of anticipating future events.

Psychokinesis

Is there such a thing as mind over matter? In a sense, the answer is easily demonstrated. If we assume that mind is not brain-based and is nonmaterial then it moves the body (matter) all the time, but this is not the type of mind affecting matter of interest here.

It seems that most people believe that they have some influence over physical objects or events. For example, gamblers attempt to influence the throw of dice in a game of craps or the spin of a slot machine. Nearly all golfers, including top professionals, "talk" to their ball, hoping it will listen and behave a certain way after they hit it. Belief that intention can affect the world or bring good luck is pervasive. The question is whether this is possible or just a myth. The scientific method provides the answer.

Experiments studying the effects of mental intention on the throw of dice have been going on for almost eighty years and provide conclusive evidence that such effects are real. A meta-analysis of 148 independent experiments indicated that the odds that mental intention has *no* influence on dice are statistically one in a billion.[108] When no intention is applied to the throw of dice, researchers found that the hit rate was exactly that predicted by chance. Do these results prove that mind can affect the throw of dice? One would have to conclude, based on the high quality of the scientific controls employed and the reproducibility of experiments by more than fifty investigators that the answer is yes.

Today most of the studies investigating psychokinesis involve experiments trying to affect random-number generators (RNGs).[109] Such experiments are easier to control and can involve trillions of randomly generated bits (ones and zeros), which can be recorded automatically and with ease—in fact, the whole test can be automated eliminating operator bias or error. The effect also tests mind-matter interaction on what is believed to be a quantum level—which fits the current model of how mind fundamentally affects matter. In addition, the experimental evidence suggests that a random system operating on a micro scale is easier to manipulate than a massive object like a die.

Radin and coworkers performed a meta-analysis of 152 published reports from 1959 to 1987 describing 597 studies in which persons were instructed to try to make the RNG generate an excess of either zeros or ones; and another 235 control studies in which no mental intention was used. Overall, the experimental studies generated a 51 percent hit rate while the controls were exactly 50 percent. Odds that such results would occur by chance are one in a trillion.[110]

Further evidence that human intention can influence sensitive physical devices, systems, and processes comes from PEAR. In a seven-year study ending in 1996 over 1200 studies using 108 volunteers attempting to influence random-number generators showed there was a small (about one tenth of one percent), but statistically significant effect on the output of RNGs (one in four thousand).[111] Analysis of the data indicated that the results were not due to the ability of a few outstanding performers but to a widespread ability found in the general population. The results also confirmed that distance and time delays did not lower the effectiveness of mental intention on the output of RNGs.

In another PEAR study human operators attempted to bias the output of a variety of mechanical, electronic, optical, acoustical, and fluid devices to conform to pre-stated intentions—all without recourse to any known physical influences. Under control conditions, all of these sophisticated instruments produced strictly random data, but when human operators applied intention to the devices, the experimental results showed changes that could only be attributed to the intention of their human operators.[112]

A RNG study dubbed the Global Consciousness Project is currently being directed by Roger Nelson at Princeton. It began in 1998 and utilizes a worldwide network of RNGs that collect randomly generated bits along with their time stamp and send the data to a server at Princeton. Data from this ongoing study show that there is a significant spike in nonrandomness coinciding with major global events, such as the death of Princess Diana, Y2K, 9/11, and the funeral of Pope John Paul II.[113] Many other globally important events caused significant spikes in the network of RNGs, suggesting that the collective mind of humans on Earth have a small but reproducible effect on what is known to be a random process.

Strong evidence that a person's intention can influence an electronic device comes from investigations involving Intention Imprinted Electrical Devices (IIEDs). Stanford physics professor William A. Tiller and collaborators Walter Dibble and Michael Kohane have performed hundreds of experiments investigating IIEDs. In a typical experiment, two identical pH meters might be used. One instrument is placed next to a highly qualified meditator who attempts to imprint the device in a manner that, for example, it gives low pH readings. The devices are then wrapped in aluminum foil and shipped approximately 2,000 miles to a laboratory where the actual target experiment is conducted by others. There a sample of purified water is prepared with a small amount of added potassium chloride (to increase conductivity of the water). Half the sample is put into a clean beaker, the pH electrode placed in the solution, and the test meter turned on. The other half is measured similarly with the control meter. The unimprinted meter reads a pH of 6.0 and the imprinted meter reads 5.0. The researchers repeat the experiment, but this time the meditator is asked to imprint one of the meters with a high pH. Interestingly, now the imprinted meter reads one pH unit higher than the control meter.

Using strict laboratory controls Tiller, et al also found significant effects of intention on the UV spectrum of DNA-water solutions, the thermodynamic activity of enzymes, the growth rate of fruit-fly larva, the temperature, and conductivity of water, and the breakdown voltage of gas-discharge tubes. The results are both robust and reproducible, and significant to the point that no statistical analysis of the data is required to demonstrate that the effects are real. Their conclusion is that there can be no doubt that electronic devices imprinted by experienced meditators give significantly different measurements of physical properties than non-imprinted devices.[114]

While some physicists believe that consciousness has nothing to do with wave function collapse, other scientists have actually conducted experiments that show that it does. One such study was published by Dean Radin and coworkers. In this experiment, a double-slit system was used and 685 people participated over the Internet, and were told to concentrate on the apparatus in California. Normally when photons in a double-slit experiment are observed the interference pattern disappears. In this case the interference pattern was degraded and the results were highly significant ($p=2.6 \times 10^{-6}$) compared to control runs that produced no degradation of the interference pattern.[115] And recently these results were confirmed by an independent investigator.[116] Such experiments indicate that mental intent does in fact influence wave function collapse influencing whether quanta behave as particles or waves. One wonders whether experiments like this will ever convince the physics community that mind or consciousness is actually the ingredient necessary for bringing physical reality into existence. Perhaps it would take someone with extraordinary powers of concentration that could turn the interference pattern on or off like a light switch from a remote location?

Premonition, Telepathy, or Clairvoyance?

Both telepathy and remote viewing could be attributed to precognition since the subjects are normally informed of results after the test is completed. Conceivably the positive results of such testing could be due to the knowledge they receive in the future. This illustrates the difficulty in devising experiments that unambiguously demonstrate a specific type of psi phenomena. As a result, parapsychologists are resigned to the fact that psi is a general phenomenon that can be demonstrated in almost any person and not separable into clearly defined classifications.

What are sometimes called telepathic impressions probably fall into this "unknown" category. These involve strong emotions, physical symptoms, or feelings that something is happening to another person they have close ties with, but are at a distant location. In most cases, the "target" person is in a serious situation–seriously ill, injured in an accident, close to death, or has died.

Larry Dossey in his excellent book, *One Mind*, provides numerous examples of telepathic impressions. One such case was a middle-age woman who developed a powerful feeling that her son who lived thousands of miles away was in serious jeopardy. She tried to ignore the feeling but it only grew stronger. Then a series of numbers popped into her mind and she developed the urge to dial the numbers, which connected her with the emergency room of a hospital in the city where her son lived. When she explained to the ER nurse that she was concerned about her son, she was told that her son was seriously injured in an automobile accident but his injuries were not life threatening.[117]

Ian Stevenson (1918-2007) was a psychiatrist who founded the University of Virginia School of Medicine Division of Perceptual Studies. His principle interest was investigating reincarnation and paranormal phenomena. During his over forty years of extensive field research he investigated many people who reported that they had a strong telepathic impression about the fate of a loved one that later turned out to be accurate. In 1970, he published a review of 160 such cases.[118] For example, a man develops a severe pain in the temple at about the same time that a relative shoots himself in the temple but before news of the event reaches him.

There have been many such reports of people encountering life-changing events that happen to loved ones before information of the event has reached them by normal means. In addition, many people have reported that because of a premonition they avoided death by refusing to fly on an airplane that later crashed.[119]

Answering the Skeptics

Materialists are naturally skeptical that psi phenomena are real. The existence of psi negates the physicalist's doctrine that mind is localized in the brain. One of their arguments is that psi would violate the laws of physics. Take for example the following statement from Sean Carroll, physics professor at the California Institute of Technology:

The only problem is parapsychology is not science. It's pseudoscience. From a completely blank-slate perspective, one can certainly pose scientific questions about whether the human mind can tell the future or read minds or move objects around without touching them. The thing is, we know the answer: no. The possibilities have been investigated and found wanting; more straightforwardly, they would violate the known laws of physics.[120]

There are several problems with such an argument. First, in this quote Carroll is spouting scientism not science. Because Carroll has an emotional attachment to a materialistic worldview—he won't even consider the possibility of a nonmaterial reality. He has no interest in even looking at the data that establishes the factual nature of psi and instead attempts to trivialize it by calling it fake science and saying it violates physical laws. He fails to recognize that he bases these statements on *belief* rather than science. A belief system (an example being a religion) has no place in science since it is unfalsifiable. Adherents to such a system accept it as factual and are not interested in investigating any evidence that might disprove it or cause them to modify their preconceived ideas of reality. Scientism is insidious because it disguises itself as science when in reality it is merely an attempt to use scientific jargon to promote one's personal point of view.

Secondly, Carroll is an advocate of the many-worlds interpretation of quantum mechanics, which postulates the reality of a universal wave function. Such a wave function would describe a universe in which everything is entangled beginning with the Big Bang. By extension if all physical reality is entangled there is every reason to assume that minds are entangled.

Thirdly, it is a scientific fact that intention, observation, measurement, information, mind, consciousness—or whatever you want to call it—has an effect on quantum systems and thus the unfolding of physical reality. The observer effect of quantum mechanics might be called "micropsychokinesis." Hence, if mind is nonphysical, nevertheless it could influence the outcome of a physical system. Bending a finger might be an example. Intention to do this might cause the appropriate neurons in the quantum mechanical brain to undergo wave function collapse in such a manner that specific muscles contract.[121]

Finally, although Carroll is an outspoken advocate of materialism, he fails to acknowledge that physical laws are not material. In fact, it is now recognized that all physical laws are fundamentally probabilistic, and therefore indicate tendencies—not firm reality. It might be unlikely that an electron placed in a metal box will suddenly be found in a neighboring galaxy, but it is nonetheless possible. Physical laws describe overall tendencies for things to happen on the macro scale—but on the microscale, things are much fuzzier and unusual events occur all the time. As a result, the behavior of a random physical process is not defined by a single event but by the combined behavior of the entire system. Over a short period, even a random process may appear nonrandom. For example, if a coin were flipped a thousand times it would be expected for heads to fall ten times in a row.

Even such great scientific minds as Albert Einstein argued that quantum entanglement violated physical law when in fact it has been proven to occur; and our current understanding of the domain of the wave function does a good job of explaining it. So it should come as no surprise that some scientists make the same argument about psi phenomena when, in fact, such phenomena fall completely within the framework of the known physical laws governing the behavior of the micro world.

Scientists who do psi research apply their time and talent to their work not because of any preconceived idea about reality but because of a real scientific curiosity, and they are not immune to criticism from those who think all such investigations are phony. As a result, they typically take extraordinary care to maintain the highest scientific controls and procedures. Unlike their colleagues in other areas of psychological research, they have had to answer a number of criticisms of their research, such as the following.

- *The studies are not well controlled. If better controls were used then only chance results would be obtained.* Actually, the data is contrary to this. As studies were increasingly controlled and refined in face of criticisms, the results did not become less significant but actually increased in statistical significance vs. random chance.[122] In addition, some researchers, being sensitive to such criticisms, consulted with professional mentalists and magicians who could find no flaws in their methodology. Interestingly, it has been shown repeatedly that skeptics of psi do worse in tests than people who believe or are at least open to the idea.
- *The studies are not reproducible.* Actually, the studies are reproducible. Many studies have been independently replicated. For

example, presentiment studies were replicated by an independent investigator using his own equipment.[123] Moreover, there have been hundreds of ganzfeld telepathy experiments, all getting positive results.
- *Nonconforming data is not reported.* This is the so-called "waste-basket" hypothesis that only studies showing positive results are published, which creates bias in the data. If the unpublished negative results were included in the data then the meta-analysis would not show positive results. The problem is the statistics do not support this criticism. For example, there would have to be fifteen unreported trials showing negative results for each reported ganzfeld study. For the remote viewing card experiments, it would take about thirty thousand unsuccessful studies in addition to the thirty-four reported studies to reduce the odds to chance.[124] Hence, considering the time and effort it takes to conduct even one study, selective reporting cannot possibly explain the positive results obtained for psi phenomena.
- *Results are fraudulent or faked.* Investigators really want to believe that psi is real and either consciously or unconsciously manipulate the data in order to achieve their expectations. While there have been a few cases of fraud or error, this is no reason to dismiss the vast majority of data. And while it is true that skeptics often get poorer or negative results when they attempt to replicate experiments of scientists who study psi phenomena,[125] accusations that seasoned scientists, most of whom work at well-respected centers of higher education, would deliberately misreport their findings is patently ridiculous. Such scientists are fully aware of the controversial nature of their research and have every reason to want to protect their professional reputations. Therefore, they take a rigorous approach to insure that their work is of the highest standard possible—much more so than typical psychological studies. Furthermore, even if one or two studies out of two hundred were biased or flawed, it would not explain the other 198 studies, which gave positive results. In addition, statistical outliers are typically not included in a data set undergoing meta-analysis.
- *Science is constantly explaining the supernatural in natural terms.* Throughout history, science has discovered that many phenomena that were earlier thought to be paranormal are actually explicable in natural terms. Of course, this is a phony argument since it says

nothing about whether psi is real, only that other phenomena, originally thought to be mysterious were later explained by science.
- *Psi effects are tiny, irreproducible, and can only be shown to occur using statistics.* This is an invalid criticism since the same can be said of most studies in the "soft" sciences. In addition, results are reproducible and in some studies, e.g. using IIEDs the effect is so robust that statistics are not required to demonstrate the scientific validity of the effect.
- *There is no theory to explain psi.* This is untrue. Psi phenomena can be explained by the existence of cosmic mind, which can also be called the one mind or collective unconscious or mental internet. These are names used to describe a collective mind, which is nonlocal in nature and represents the wholeness that is a fundamental property of the domain from which physical reality emerges. The nature of cosmic mind is discussed in Part Three of this book.

Summary

Much of the skepticism of psi phenomena falls in the category of pseudoskepticism, which is a type of false skepticism where one pretends to have a questioning attitude, but in reality, one's mind is hermetically sealed against whatever may threaten their preconceived idea of what is true. A pseudoskeptic is not an expert in the field they are trying to debunk. Rather they are a believer in an alternate system and spout scientism. They only try to appear open-minded and scientific when they are neither. A true skeptic must be genuinely curious enough to look at the data from an agnostic point of view before rendering an opinion. A true skeptic is one who acknowledges that the burden of proof is not solely on those who make extraordinary claims but also on anyone who claims that the mountain of evidence supporting a hypothesis is incorrect for some reason. The critic is also making a claim and must bear a burden of proof. They have a duty to suggest a way to overcome any deficiencies they claim are in the data. Pseudoskeptics fail to recognize these truths. Typically, they see no reason even to consider the evidence since they are convinced that psi is impossible.

In truth, any person that looked at the evidence for psi would have to admit that psychic phenomena have now been established as scientific fact. Just as the simplest components of physical reality can be entangled, we now know that minds can be entangled. This truth should provide us with great confidence that materialism is a false ideology.

13

EVIDENCE FOR LIFE AFTER DEATH

Throughout human history, even among diverse cultures, there has been an almost universal belief in some form of afterlife. Some of the earliest examples of this are from archeological evidence of Stone Age burial rituals. Implicit in such burial and funeral practices is the assumption that death is not the end, but it marks a transition. Other examples are those of native peoples and ancient civilizations such as the Greeks and Egyptians. Moreover, almost all the folk religions of yore and today's major religions proclaim that our existence does not end with death. The question is whether the scientific method can provide any evidence for the continuation of consciousness or life after death.

We have already seen some evidence for the continuation of consciousness after the cessation of brain activity from people's anecdotal descriptions of near-death and out-of-body experiences. They consistently describe having fully functioning, lucid consciousness, and self-awareness while witnessing their lifeless body from above. This implies that mind and consciousness can function independently of the brain/body and therefore could survive death. The concept that a part of our being can survive death is consistent with the idea that mind is nonphysical but normally depends on a functioning brain to express itself in the realm of physical reality.

The incorporeal essence of our being that can function independent of brain, and theoretically continues after death can be called the soul. It would be the part of our being that survives death, and it could carry memories, personality, and karma from this life to the afterlife—whatever that is. However, the term soul has religious overtones and is not well defined by the various religions. For this reason, I choose to call it "unit mind."

Additional scientific evidence for the existence of unit mind and its continuation following death of the body comes from children and adults that have detailed and verifiable memories of their past lives, and from people that display exceptional talents and genius that defy rationality. In other words, there is a considerable body of the scientific evidence for life after death that comes from studies that suggest that reincarnation is the mode in which the nonphysical aspect of our being survives death.

Mention the word reincarnation to most westerners and they will probably think that it is weird and esoteric—something that Hindus and Buddhists believe in. But is this really the case?

Is the Idea of Reincarnation Really that Weird?

Reincarnation is defined as the transmigration of an individual's unit mind from one body to a new body following death. It is a fundamental tenet of all Eastern religions (Hinduism, Buddhism, Taoism, Jainism, Sikhism, etc.). However, it was also a tenet of the ancient Egyptians, Greeks, Celts of Great Britain, Vikings, folk religions of Africa, Australia, East Asia, Siberia, shamanism of South America, and the indigenous Oceanic peoples. In addition, Native Americans have a long history of belief in reincarnation, and today the Iñupiat people of northwest Alaska, the Aleuts of the Aleutian Islands, and the Tlingit Indians of Canada maintain this belief. In addition, Judaism, Islam, and Christianity have deep ties to this doctrine; and to this day, the Kabbalist and Hasidic sects of Judaism and the Sufi, Ismaili, and Druze sects of Islam accept and teach the doctrine.

The concept of reincarnation is closely entwined with the idea of karma. The law of karma states, "we reap what we sow." Our karma is the sum of our past actions, both good and bad, that we have retained in our unit mind but in a sense have not yet experienced or "burned." We are born into circumstances that allow us to experience life in such a manner that we can best burn through our remaining karma and progress toward the ultimate goal of union with God. In other words, our karma helps us learn important life lessons, including how some actions bring us happiness and others result in pain and suffering. Through this process, we grow in wisdom and move forward on the path to spiritual union.

During the first five hundred years of its existence, Christianity had a strong connection with reincarnation as evidenced by the beliefs of Gnostic Christians, the teachings of Origen, and other Neoplatonists. Some have argued that Jesus taught reincarnation but that the Church both misinterpreted and suppressed his teachings of it.[126,127,128,129]

Today belief in reincarnation is common among the world's population, and by some estimates the number of people who believe in reincarnation exceed the number who reject the idea or have no knowledge of it. Recent surveys showed that 20 percent of Americans believe in reincarnation and 26 percent of Canadians, while for Europeans the number was close to 30 percent.[130] The number of Westerners that believe in reincarnation might seem surprising considering the fact that it is not a matter of western doctrine or faith.

Numerous prominent personalities have stated that they believe in reincarnation. A list of Americans includes Benjamin Franklin, Ralph Waldo Emerson, Henry David Thoreau, Mark Twain, Thomas Edison, Henry Ford, Walt Whitman, Eleanor Roosevelt, General George S. Patton, Willie Nelson, Shirley MacLaine, Tina Turner, Wayne Dyer, Richard Gere, John Travolta, and Oprah Winfrey.

Children Who Remember Their Previous Life

Most of the evidence confirming reincarnation is in the personal accounts of individuals who have provided detailed and factually accurate descriptions of their previous life. Probably the majority of such accounts are from children who spontaneously describe their previous life to their parents in such detail that the person described can be identified and their life story checked against the facts provided by the child. There have also been extensive studies of cases suggestive of reincarnation where birthmarks and birth defects can be traced to the circumstances of the death or to the personal history of the person identified as their predecessor. Finally, there are numerous accounts of adults remembering previous lives—most often, under what is termed "regression hypnosis."

One of the pioneers in such studies was Ian Stevenson who was mentioned earlier and worked at the University of Virginia School of Medicine Division of Perceptual Studies. He traveled extensively over a period of forty years, investigating three thousand cases of children around the world who told their parents they remembered their past lives.

His first book on the subject was *Twenty Cases Suggestive of Reincarnation*.[131] Half the cases were of children living in India and Sri Lanka where reincarnation is culturally accepted as fact, while the other ten cases were from Brazil, Lebanon, and from the Tlingit Indians of the southeastern coast of Alaska. A case usually started when a small child of two to four years of age began talking to their parents or siblings of a life they led in another time and place. The child usually felt a considerable pull toward the events of their previous life and they frequently pestered their parents to let them return to the community where they felt they formerly lived. If the child made enough specific statements about their previous life, Stevenson would verify that their memories accurately recalled details from the life of a deceased person that the child had identified.

In a second book entitled *Children Who Remember Previous Lives: A Question of Reincarnation*,[132] Stevenson described fourteen cases typical of children who remembered previous lives, detailed how such studies were conducted, cultural influences, and how reincarnation might explain many of the unusual abilities, inclinations, behaviors, and abnormalities in children that have no known cause.

Undoubtedly, Stevenson's greatest contribution to research on reincarnation was his two-volume, 2268-page publication *Reincarnation and Biology: A Contribution to the Etiology of Birthmarks and Birth Defects*.[133] The study contained 225 case reports of children who remembered previous lives and who had physical abnormalities that matched those of persons described by the child. Many of his subjects had unusual birthmarks and birth defects, such as finger deformities, underdeveloped ears, or being born without a foot. There were also scar-like, hypopigmented birthmarks and port-wine stains. His studies detailed cases of birthmarks and birth defects in children that corresponded to a wound (often fatal) on the deceased person whose life the child recalled. These cases were normally confirmed by autopsy findings and photos.

An example of a case studied by Stevenson goes as follows. An Indian boy was born with a large hypopigmented birthmark on his chest. When the boy was four years old, he began to talk about being a man in his previous life that was murdered in a nearby village by a shotgun blast to the chest. Stevenson interviewed the boy and his mother and was able to identify the man the boy described. Then he was able to find a close similarity between the boy's birthmark and the postmortem report of the shooting victim.

Stevenson was an expert on psychosomatic medicine and suspected that strong emotions like a traumatic death may be related to a child's retention of past-life memories. Many of the children he studied claimed that they had met a violent end in their previous life and sometimes the child's fears and abnormal behavior could be linked to the way they died. For example, a girl who claimed to have drowned in her previous life might have a fear of water, while a boy who died from a fall from a height might display a fear of heights. He came to believe that neither environment nor heredity could account for certain fears, illnesses, and special abilities, and that some form of personality or memory transfer would provide an explanation for these. Stevenson came to believe that almost everybody has experienced past lives, but only a small percentage of children retain any memories of their previous life.

After working several years with Ian Stevenson at the University of Virginia School of Medicine, Jim B. Tucker, MD has taken over leadership of the Division of Perceptual Studies. He has authored two books on children's memories of previous lives: *Life Before Life* and *Return to Life*.[134,135] Like his mentor, Ian Stevenson, his research has primarily been of children that remember their previous life and have provided enough details that the person the child described can be identified.

An example would be the case of a boy named Ryan Hammons who lived in Oklahoma. At the age of four, Ryan began describing his vivid dreams of his previous life as a Hollywood agent. Ryan provided many details of his previous life, such as being married five times, having one daughter, two sisters, meeting actor Rita Hayworth, taking trips to Paris, dancing on Broadway, and living on a street with the name rock in it. His mother said his stories were detailed and extensive, unlike something a child would make up. One day, when going through some old Hollywood picture books, Ryan immediately identified himself in a picture. However, the picture did not name the actor, and Ryan's mother could not find any more information about the man.

After hearing of the work of Jim Tucker, she approached him for help in identifying the man in the picture. After considerable research, Tucker was able to identify the man as Marty Martyn and confirm that he had been a bit actor turned Hollywood agent. Numerous details about Martyn's life that the boy had provided checked out. For example, Martyn had danced on Broadway, traveled to Paris, worked with Rita Hayworth, been married five times, had two sisters, one daughter, and lived on Roxbury

Drive. Indeed most of the details about Martyn's life and professional career as a Hollywood agent (mainly supplied by his daughter) checked out with the facts supplied by Ryan. In all Tucker was able to confirm fifty-five details provided by Ryan that accurately described Martyn's life. The case even attracted national attention.[136]

To date researchers at the University of Virginia Medical School's Division of Perceptual Studies have compiled over 2500 reports of children from around the world who claim to remember their past lives. The data obtained by these researchers provide convincing evidence for reincarnation. They provide solid evidence that mind and consciousness survive death.

Past-life Regressions

Past-life regression is a technique that uses hypnosis to recover what practitioners believe are memories of past lives. Past-life regression is typically undertaken in pursuit of a psychotherapeutic cure—particularly of irrational fears (phobias). Although remembrance of past lives may occur spontaneously in some adults, it is much more common when undergoing hypnotic regression. As it turns out, almost anybody who can be deeply hypnotized can be regressed by a hypnotist and relate their experiences of past lives. The experiences are most often visual and take on a life of their own without intercession by the therapist. Subjects identify strongly with a particular individual and feel the emotions of that person during the regression. Often these feelings mirror problems faced by the subject in their current life. Finally, most of the subjects feel like a weight has been lifted from their mind after they relive a particularly traumatic experience from a previous life.

Dr. Raymond Moody has been a prominent researcher of past-life regressions. He began as a confirmed skeptic with regard to reincarnation. But as he began to investigate both psychologically healthy patients and those with phobias who under deep hypnosis suddenly began describing in vivid detail episodes from other historical periods they could not possibly have known, his interest in the meaning of these visions intensified. He found that many of his patients were cured of their phobias when they relived a traumatic event from a previous lifetime.[137] Questioning what his patients were describing he decided to undertake his own journey into the mysterious realm of past-life experiences. He performed self-hypnosis and became aware of nine of his former lives.

Another researcher into past-life regressions is Dr. Brian Weiss. He was a traditional psychiatrist who discovered that some of his patients recovered from chronic neuroses following past-life regressions in which they described events related to their current fears and anxieties. Initially he was a skeptic about reincarnation, but his skepticism was gradually eroded when he was able to confirm elements of his patient's stories through public records and was able to cure their neuroses. Today Weiss is convinced that elements of human personality survive death and that many phobias and ailments are rooted in past-life experiences that when relived by the patients have a curative effect.[138]

Currently one of the best-known practitioners of past-life regressions is Carol Bowman. She has done thousands of past-life regressions in her practice. She was first introduced to past-life regression in 1987 while living in Asheville, NC. She underwent a regression to see if she could identify the cause of her chronic lung problems. In just one session, she experienced two lifetimes in which she had died due to afflictions of the lungs—dying of consumption in the nineteenth century and then in the gas chambers of World War II. That one session changed the course of her illness and convinced her that we really do live more than once. Bowman has written two books on reincarnation. In her first book, *Children's Past Lives*, she revealed overwhelming evidence of past-life memories in children; showed how young children remember their past lives—spontaneously and naturally; and explained that such memories are far more common than most people realize.[139] In her second book, *Return from Heaven*, she describes her research into cases of reincarnation within the same family.[140] She documents the emotions and relief that families experience when they discover that a deceased family member has returned in a new body as a child in the family.

Bowman currently oversees a popular website called the Reincarnation Forum where people can post their reincarnation and past-life experiences.[141] The Reincarnation Forum currently has over ninety thousand posts.

Other Evidence Suggestive of Reincarnation

Xenoglossy is the ability to speak or write a language without having learned it. Stevenson reported a handful of cases of xenoglossy, including two where subjects under hypnosis allegedly conversed with people speaking a foreign language they did not know.

The medical literature has reports of people with dissociative identity disorder (multiple personality disorder) where one of the personalities can speak a foreign language fluently that they had not previously studied. One such case was that of an Indian woman named Uttara who grew up in Nagpur, India speaking the Marathi language. She had an unremarkable life as a teacher at the local college until at the age of thirty-two she began to undergo personality changes and at regular intervals would take on a second personality. In this dissociated personality, she was Sharada, a Bengali woman that lived in the 1800s. When the Sharada personality took over, Uttara could speak fluent Bengali, much as it would have been spoken in the early nineteenth century, yet Uttara had only a cursory knowledge of Bengali before she began suffering from the disorder.[142]

Déjà vu is French for "already seen." It is the experience or feeling that one has lived through the present situation before, or the feeling of familiarity with a place that one has never visited. Reincarnation is a possible explanation for the déjà vu experiences of some people. They may know their way around a locality that they are visiting for the first time. The whole place, or at least a significant part of it, may seem familiar to them. Reincarnation is not the only explanation for this common experience. A clairvoyant vision or precognitive knowledge could be involved.

Reincarnation might also explain how some children develop unusual interests, extraordinary skills, abilities, and genius—often in contrast to the expectations of their family. Stevenson reported on a number of such cases. A young Bengali girl began to produce elaborate songs and dances without prior training; a young Indian boy started playing the classical drums or *tablas* with great skill; and another child displayed unusual aptitude working with marine engines.[143]

Some children are considered prodigies because they display expertise in some activity or area of knowledge; have the ability to speak with emotional clarity and maturity that greatly surpasses their life experience; have extraordinary presence and awareness; have the ability to focus during a conversation or on a topic without the restlessness typical of other children their age; have exceptional memory and/or the ability to learn things very quickly and develop a deep understanding of a subject; and have ease in communicating with adults. In other words, they display knowledge and behaviors that would be consistent with an individual that was much older.

Examples of children's prodigious abilities include hyperlexia (the ability to read at a very early age without prior training), extraordinary memory, and exceptional talent for music, math, art, language, and mechanical domains. Some of these children are given the label of savant. This syndrome is a loose term that refers to people who have a combination of significant cognitive difficulties, often attributed to autism, and profound skills. An example would be Kim Peek, who was the inspiration for the main character in the 1988 Oscar-winning movie Rain Man. Peek could read a book an hour, memorizing two pages at a time (left eye reading the left page, right eye reading the right), and afterward he was able to reproduce 98 percent of the content by heart. Still other savants have a facility with foreign languages, ability to draw landscapes from memory, amazing skills with numbers or math, the ability to measure distances or heights with precision (without instruments), or exceptional map-reading skills.

Genius is another type of an extraordinary ability that often appears to arise from nowhere. For example, Isaac Newton was not born into a family with any interest in mathematics, nor did his three siblings show any interest or ability in this area. Interestingly he was born one year after the death of Galileo and his life's work was a continuation of that of his predecessor.

Mozart displayed prodigious musical ability from his childhood. Already competent on keyboard and violin, he composed music from the age of five, and while still a boy, performed before European royalty. Mozart received only rudimentary training in music from his father whom he surpassed in musical composition at an early age.

Most would agree that Albert Einstein was a genius, but where his genius came from is anyone's guess. His parents were very nurturing and encouraged him to be independent and creative, not only in science but also in music, paying for piano and violin lessons. However, neither of his parents, nor his two sisters, nor his three children showed any significant aptitude for mathematics.

Scientists cannot currently explain why some children seem to be born with an exceptional knowledge, talent, ability, or interest when there is nothing in their nurturing or genes to explain it. Yet throughout history, we can find numerous examples of individuals, who from a very early age had a predisposition, motivation, or ability that was extraordinary or a result of genius. If indeed, we are solely a product of our genes and upbringing—that is the materialist model—then where do these exceptional abilities come from? A more reasonable hypothesis is that the body is only temporary but the unit mind is permanent, and it can carry knowledge, abilities, and inclinations from previous lives to the current life.

The How and Why of Reincarnation

Assuming mind is nonphysical, it clearly needs a healthy brain to express itself in the physical world. Under certain circumstances such as cardiac arrest and shutting down of the brain, the mind appears to be able to leave the body, yet retain full conscious awareness. Upon death, the unit mind without a body is termed a "bodiless mind." Such a bodiless mind could retain memories, personality traits, and karma from its previous life. However, lacking a physical body it would be unable to affect anything in the physical realm.

Spiritual sages throughout history describe reincarnation as the process by which the bodiless unit mind remains suspended in a transition state similar to deep sleep until it finds a new home in a fertilized ovum that will best allow it to continue on its spiritual journey toward eventual union with cosmic consciousness. In addition, they tell us that our actions bring reactions. These reactions can be good or bad, but some reactions are not experienced immediately but are stored in the mind to be experienced later. These stored reactions constitute our karma and since we earned our karma in a physical body, we can only experience it in this or a future physical body. Hence, we might require more than one lifetime to attain union with God.

Assuming that God is just, we might agree that reincarnation is a logically more satisfying concept of the afterlife than the alternative proposed by Christianity. The Christian doctrine of the afterlife can largely be attributed to Augustine (354-430), whose Manichean worldview left an indelible stamp on Christianity.[144] In a nutshell, the doctrine states that the soul comes into existence at the time of conception, and it is an accident of nature if an innocent child is born blind or without arms while another is born a prince. The reason that God allows this to happen is because man is tainted by original sin, and the fate of a child is left to chance because its ancestors committed a crime against God. Furthermore, only by accepting Jesus into our life can we experience salvation. This entails bodily resurrection and an unending afterlife in a heaven devoid of suffering. If salvation is not attained in a single lifetime, Augustine argued that we would suffer eternal damnation. The only problem is that the Old Testament never mentions the existence of a realm for the good (heaven) and a realm for the evil (hell); and in fact, it never claims that the souls of people continue to live in any form after they die.

Summary

The title of this chapter suggests that there is evidence for life after death. The most important requirement for life after death is that mind is non-physical, and it can continue to exist when the brain ceases to function. We have already seen the overwhelming evidence that mind cannot be reduced to brain. This model does not imply mind-body dualism, but does require that there is a hierarchy of consciousness > mind > matter. There actually is no dualism—just a change in phase—ultimately everything is composed of consciousness.

The scientific evidence for reincarnation comes primarily from adults and children who remember their previous lives, and from the inexplicable emergence in some individuals of unusual interests, extraordinary skills, abilities, and genius. Most of the evidence from adults remembering their previous lives comes from hypnotic regressions. Skeptics of reincarnation are quick to attribute this to cryptomnesia (a forgotten memory that returns without its being recognized as such). However, such an explanation does not work when it comes to children who lack extensive life experiences and could not possibly know details about the person they are describing. Based on the high quality of the evidence that some children have veridical memories of past lives, and the fact that many of these children also bear birthmarks or birth defects consistent with those of the person they identify with, one would have to conclude that there is strong evidence for reincarnation and the continuation of consciousness following death. As renowned skeptic, Carl Sagan put it, "Young children sometimes report the details of a previous life, which upon checking turn out to be accurate and which they could not have known about in any other way than reincarnation."

Of course, reincarnation is not the only possible explanation for children's memory of previous lives. For example, such knowledge could be attributed to ESP. However, it is not clear how this would explain the birthmarks and birth defects seen in such children or the personality traits and inclinations that are often observed in children that mirror those of the person they identify with in a previous life. In any case, ESP is not consistent with the materialist worldview and skeptics would reject this as an explanation for reincarnation memories.

Overall, we conclude that life after death, which is consistent with the spiritual worldview, is supported by strong scientific evidence and a mountain of anecdotal reports. This should increase our credence that this worldview is correct and that the materialist worldview is false.

14

FALSIFYING THE MATERIALIST WORLDVIEW

How Quantum Mechanics Challenges Materialism

WE HAVE SEEN HOW the materialist worldview has had to undergo modifications in light of the discoveries of quantum mechanics and relativity theory. Relativity theory never posed a direct challenge to materialism because it was still deterministic and local; but it only described the behavior of macroscopic objects. Quantum mechanics on the other hand has had a much more devastating effect on this ideology. Things first started to unravel for the classical materialist model of reality when unexplained anomalies arose concerning the spectrum of hydrogen and the quantized nature of black-body radiation. Eventually quantum mechanics was developed by physicists that explained these anomalies plus many other qualities of the micro world. Its success in doing this has been no less than spectacular, and some day scientists may come to a consensus that quantum mechanics is the greatest scientific discovery of all time.

The two greatest challenges that quantum mechanics poses to materialism are: (1) nonlocality—quanta can be entangled or connected no matter how far apart they are; and (2) a quantum behaves differently if we look vs. we don't look.

As we have seen, the development of the wave function with its mathematical description for the possible states a quantum system might assume when looked at provided a huge boost to understanding the behavior of quanta. Nonlocality can be explained because quanta emerge from a subtler or more fundamental domain—that of the wave function. In this domain, there are no separate parts. Everything is entangled or connected, and it is not located in any place in space but is everywhere. Materialism copes with this description of reality by saying that matter may not be the ground substance of creation, but that it is an emergent property.

However, early in the development of quantum mechanics, physicists discovered that quanta appear to know whether we are looking at them. When we are not looking, they go on happily behaving like waves, but when we look, they revert to their particle nature. Since no one believes they have a brain and could sense whether they are being watched, physicists had to come up with an explanation for this weird behavior. The problem for materialism was the part about the need for quanta to be looked at. In order for matter to emerge from this more fundamental reality *someone or something needed to look at it*—that is, collapse the wave function. Since any physical instrument or apparatus that might be used to observe a quantum would also be described by a wave function, then what would collapse the wave function of the quantum? Eventually we need something *nonphysical* to cause such collapse. In the absence of such a nonphysical aspect of reality, theoretically nothing could emerge from the realm of potentiality into physical reality. For this reason, many of the pioneers of quantum mechanics believed consciousness was needed for the collapsing. However, materialists would have none of this. According to their model—mind and consciousness are epiphenomena of matter—they are *material* in origin.

Despite the difficulty of explaining wave function collapse without abandoning the materialist ontology, most scientists today remain firmly in the materialist camp. They just don't want to concern themselves with interpretations of quantum mechanics and the implications of the theory. Those scientists that are concerned with what quantum mechanics implies about reality are often attracted to the many-worlds interpretation (MWI) because it seemingly does not require someone or something to look at quanta before they emerge into physical reality. In fact, all possible states emerge from the subtle domain of the wave function—we are just not aware of them.

MWI postulates that conscious beings are not required to initiate splitting of the wave function—only "measurement-like interactions."

Only problem is that these interactions are not defined nor is the theory testable or falsifiable. A theory or explanation that cannot be falsified falls outside the domain of science. Any such theory that purports to be scientific but is in fact immune to falsification is a classic example of pseudoscience.

Finally, experiments prove that it is not measurements that cause wave function branching (or collapse) but *information* about the measurement, and this information may be obtained retroactively. If measurement-like interactions are occurring all the time in the universe then exactly who is "informed" about them? This need to include the role information plays in MWI branching implies that the universal wave function postulated by the theory has a *mental component*; and this is exactly the argument of spirituality.

Another problem with the many-worlds theory is that we have been split into countless worlds with their separate realities, but we are aware of only this one world. The existence of other worlds, if they were to exist, is irrelevant—we can only deal with the world we know. What this theory may actually be describing are the myriad possibilities that exist for us depending on our choices. As we make a choice, we go down one path. A different choice would lead us down another, and so forth. A choice takes us down one of the possible branches predicted by the MWI of quantum mechanics—but we only experience a single reality since it depends on the choice we make. Accordingly, there is no need for multiple realities. However, MWI is thoroughly materialistic and logically we are nothing but a bag of particles—free will is impossible. Hence, choice doesn't enter the MWI equation.

As mentioned earlier an interesting consequence of the MWI is that it elevates entanglement to include everything within physical reality. Ultimate reality is the wave function for the universe. This implies that there is a way of describing all the possible (allowed) states and interactions for all the constituents of the universe. Everything in the universe is entangled, and necessarily all separate parts of the universe emerge from this subtle realm of oneness. *This is precisely the argument of the spiritual worldview.* In other words this interpretation of quantum mechanics, which is favored by many materialists because it seems to eliminate consciousness from the equation, actually hypothesizes that the universe is a singularity—an indivisible unbroken whole.

Sean Carroll, physics professor at Caltech and outspoken materialist (or as he prefers to call himself: "poetic naturalist") is a major proponent of this interpretation. In his book, *The Big Picture* he writes:

> When exactly does the wave function collapse take place? (so you're not kept in suspense, almost no modern physicist thinks that "consciousness" has anything whatsoever to do with quantum mechanics. There are an iconoclastic few who do, but it's a tiny minority, unrepresentative of the mainstream).[145]

However, Carroll would be wrong in his statement. A list (in alphabetical order) of some of the "radical" and better-known physicists that have stated they think consciousness has everything to do with quantum mechanics includes the following physicists:[146]

> Edmond Bauer, John Bell, David Bohm, Neils Bohr, Fritjof Capra, Jean Charon, Freeman Dyson, Norman Friedman, Amit Goswami, Werner Heisenberg, James Jeans, Brian Josephson, Menas Kafatos, Andrei Linde, Fritz London, Henry Margenau, John von Neumann, Wolfgang Pauli, Matej Pavšič, Roger Penrose, Max Planck, Ilya Prigogine, Alastair Rae, Erwin Schrödinger, Henry Stapp, John Wheeler, Eugene Wigner, and Anton Zeilinger.

One can certainly be a renowned physicist and still believe in the materialist worldview. It's just that the observation problem is a significant anomaly that has not been adequately addressed by materialism. Secondly, everything we know about the domain of the wave function indicates that it represents a more fundamental level of reality that is timeless, holistic, and what we call physical reality only manifests from it if there is someone or thing to look at it.

The Other Glitches

There is overwhelming evidence that mind cannot be reduced to brain. Even if we ignore the problem of how consciousness is generated in the brain, there is still an abundance of evidence indicating that a change in a mental state can cause a physiological change in the body. The evidence for this was outlined in Chapter 10. Consider, for example, how a hypnotized subject who is told to hold their hand over a nonexistent candle flame can develop a burning pain and a blister on their palm—just like someone might experience if they actually placed their palm above a flame. Because the hypnotist suggests there is a flame, the subject has an expectation that they will suffer a burn. We can say that the expectation is a brain function, but the effect on the skin is not a normal way in which the brain affects the body. Normally the brain-body connection works the other way—the damaged body informs the brain of a problem—and we feel pain. In order for expectation alone to damage the body then something else must be going on. It is not enough to say that there is a close body-brain connection. One must be able to provide a physical mechanism for this process. This is the crux of the problem created by equating mind with brain. No such mechanism is known and it is unclear how one could be developed that would explain how the brain causes such physical changes in the body.

Other anomalies of materialist ontology are mystical, out-of-body, and near-death experiences. These are strictly experiential and as such, the materialist can deny their validity. However, the experiences are universally transformative and bear remarkable similarities. In addition, there is abundant evidence that lucid consciousness continues when for all practical purposes the brain is shut down due to a lack of blood flow. In addition, there are blind persons providing accurate accounts of visual experiences, and several other phenomena associated with near-death that cannot be explained in terms of current theories of how the brain functions.

The evidence that conscious experience continues after the brain ceases to function comes also from the extensive research into people's memories of living in a previous body. The high quality of the scientific research of Ian Stevenson, Jim Tucker, and others into children's accurate descriptions of living on Earth in an earlier incarnation, many of whom have birthmarks or defects consistent with the person they identify with, cannot be dismissed without providing another reasonable explanation for these phenomena. Lacking such an explanation materialists are forced to ignore the evidence, or label it as fraudulent or inconclusive.

Psi phenomena put the final nail in the coffin of the materialist worldview. At the beginning of this part of the book, I asked the question of what would falsify materialism. One would be proof that communication is possible in ways that do not use the physical senses or processes. A fundamental postulate of materialism is that mind and consciousness are generated by the brain and therefore must be local capabilities. Psi phenomena are nonlocal capabilities that cannot be explained using local field theories because they are not diminished by either distance or electromagnetic shielding. Yet, there has been over 140 years of research into psi that has provided overwhelming scientific evidence for its reality. By any measure of what constitutes scientific proof, psi phenomena meet the standard. No amount of hand waving saying that parapsychology is merely pseudoscience will change the fact that when the scientific method has been applied to the study of these phenomena the results have been consistently positive.

Materialists and skeptics are often heard to say that extraordinary claims require extraordinary evidence. However, there is nothing extraordinary about psi. It is a natural consequence of spiritual ideology and the existence of cosmic mind. What is indeed extraordinary is ignoring a huge body of high quality research because one has a dogmatic belief that nature simply does not work that way.

Summary

From quantum mechanics to psi phenomena, the materialist worldview suffers from anomalies that should make any person devoted to this concept of reality question its veracity. Scientists now accept nonlocality in the physical world as a given. There is an obvious clue here that suggests that mind, which is subtler than matter, might be nonlocal. Secondly, science discovered that physical reality is not fundamental, but emerges from the subtle domain of the wave function. Properties of this realm are timelessness, wholeness, and entanglement. There are no separate parts of reality in this realm. This implies that our experience of reality with its separate parts is illusory. Shouldn't today's materialists take note of this fact also? Materialism is based on reductionism. Everything can be explained by reducing things to the dance of the smallest parts that make things up. But what happens when the smallest part is the whole? This brings us back to a fundamental concept of spirituality.

One can still be devoted to materialism while acknowledging that quantum mechanics is weird and mysterious and has implications that are inconsistent with that worldview. However, it is not acceptable to ignore the scientific evidence that indicates mind is nonlocal and nonphysical. Anyone who was to look at the enormous body of evidence with an open mind, would have to conclude that materialism has now been scientifically proven to be false. In Chapter 23, we will consider why many scientists and intellectuals doggedly remain convinced of a false ideology in the face of overwhelming contrary evidence.

PART THREE

THE LOGIC OF THE SPIRITUAL WORLDVIEW

Any description of reality must begin with assumptions about the fundamental "ground substance" for creation. The starting material for creation might exist in actual or potential form but it would need to be eternal with no beginning or end. In materialist ontology, that substance is matter (or the domain of the wave function from which it emerges). It is meaningless to ask the question of where matter comes from. The materialist will merely shrug their shoulders and say matter exists and thus the potential for matter to emerge in the universe was always there. The question of where it comes from is unknowable and therefore meaningless. We have to start somewhere, and then build up our model of reality based on some assumptions. This is also true in mathematics where certain axioms or postulates are taken to be true, to serve as a premise or starting point for further reasoning and proofs.

For top-down spiritual ontology, the basic ground substance is taken to be cosmic consciousness. The spiritualist cannot say why or how consciousness exists, just that it has always existed with no beginning or end. Nor can the spiritualist provide a clear description of what consciousness is. It is something we possess—as do all living creatures—although it manifests in a rudimentary form in the simplest organisms. But if consciousness is defined as awareness, do rocks possess consciousness? They show no signs of awareness. How can they be composed of consciousness?[147] This outlines one of the major challenges facing spiritual ontology—explaining how cosmic consciousness is transformed into the material world. This is not a difficult problem if we assume that creation is cyclical in nature—i.e. it begins and ends with consciousness; and that consciousness has within it a creative principle that has the capability of molding consciousness into the material world.

As mentioned earlier, I am of the opinion that it is not sufficient merely to criticize the reductionist-materialist view of reality. If we are going to criticize a worldview, then we must be able to propose an alternative worldview that is both logical and rational and provides an explanation for all the phenomena that we accuse the other worldview of ignoring, denying, or being wrong about. This is what I will attempt to do in this part of the book.

What is presented could be termed the "spiritual creed." It is based on the ideas and logic that has been passed down for hundreds of generations from some of the greatest spiritual sages of the East. However, the specific model of reality I present in this section is based on the teachings and writings of Shrii Shrii Anandamurti. His was a concise, logical, and complete description of spiritual ideology. I believe that he has been humankind's greatest contemporary authority on this subject, and his model is presented in a factual manner in this part of the book.

15

HOW CONSCIOUSNESS IS TRANSFORMED INTO THE MATERIAL WORLD

The First Phase of the Cosmic Creation Cycle—From Consciousness to Rock

ACCORDING TO COSMOLOGISTS, THE universe originated from a dimensionless point, which underwent tremendously rapid expansion known as the Big Bang. Prior to this, there was nothing, including no space or time. However, cosmologists have offered little beyond pure speculation on how the enormous mass-energy of the universe emerged from complete nothingness. On the other hand, spiritual ideology claims that the universe emerged not from nothingness, but from unqualified cosmic consciousness. However, without a force or principle to qualify consciousness, nothing could be created from it. It would be like a lump of clay lacking the hands of a sculptor. Hence, according to spiritual philosophy, the Cosmic Entity, known as Brahma, has two complementary aspects—unqualified cosmic consciousness and a creative or qualifying principle. Without the action of this creative principle (CP),[148] the unqualified consciousness could not undergo any transformation and would remain unmanifest as pure awareness.

This creative principle has three basic ways or modes in which it qualifies. The subtlest mode is known as the sentient binding force.[149] When this mode of the CP acts on unqualified consciousness, it creates the cosmic idea of *I am*.[150] This is a very subtle qualification of pure awareness or consciousness, but without such an awareness of its own existence, consciousness would be nothing but pure being.

The second mode of the CP is mutative.[151] Acting on the cosmic *I am* principle, it creates the cosmic sense of *I do*.

The strongest binding force of the CP is the static mode.[152] Acting on the *I do* principle, it creates an objective sense in cosmic mind, which I call *objective mind* (OM).[153] This cosmic OM has objective reality but no physical reality, since matter is not yet formed.

The cosmic OM is somewhat analogous to a mental construct. It could be compared to an image we create in our minds. For example, suppose you close your eyes and visualize a person riding a horse. There is willfulness or doership associated with your sense of *I* as your mind takes on the colors, form, and motion of the horse and rider. For us the scene has only subjective reality, much like a dream or hallucination. It has no objective reality since we lack the power to create a living horse and rider.

These three qualities—*I am*, *I do*, and *objective mind*—are the fundamental qualities of mind. For anything we intend to do, such as to go to the store to buy a quart of milk, there is always a sense of *I* that lies behind our doership to go somewhere to accomplish our objective of obtaining milk. Similarly, cosmic mind has these three qualities. It is formed by the transformation of pure unqualified consciousness into the three principle qualities of mind: *I am*, *I do*, and *objective mind*. Cosmic mind is inherently creative and goes on to create the material world. However, without cosmic mind there could be no further creative transformation of consciousness into the material world. This is because cosmic consciousness is transcendent and unmanifest. Cosmic mind serves as the creative intermediary that allows consciousness to manifest into the material world. This first phase of the creation cycle occurs before the creation of space and time.[154]

The next phase of the creation cycle is dominated by the static principle. According to spiritual ideology, the gradually increasing binding force of the CP acting on a portion of cosmic OM transforms it into spacetime. This point in the creation cycle corresponds to the Big Bang postulated by cosmologists. While the formation of cosmic mind occurs before there is space and time, the very rapid expansion of spacetime from what could be termed the cosmic nucleus marks time zero on the cosmic time line.

Initially spacetime expanded very rapidly (much faster than the speed of light) but at the same time the pressure of the static binding principle continued to increase and very quickly it caused the expansion of spacetime to slow to the modest rate observed by astronomers today. Concurrently, the static principle transformed a portion of the spacetime into high-energy gas particles called plasma. This hot plasma emitted electromagnetic radiation (luminous factor) and eventually it cooled to give primarily hydrogen atoms with small amounts of helium and lithium.

This part of the story is the same as told by today's cosmologists. Next the extremely week force of gravity, which is a manifestation of the static CP, eventually led to hydrogen atoms being attracted to one another. Once a huge mass of hydrogen formed, the gravitational force became so strong that it caused the hydrogen atoms to compress into dense plasma (atomic nuclei without electrons). Under these conditions of extreme heat and pressure, the nuclei fused, igniting the first baby stars. This process is termed thermonuclear fusion and is the process that fuels the stars.

The earliest stars that formed were massive and as a result had short life times before they exploded in what is called a supernova. In the process heavier elements were formed which are currently being recycled by stars that formed later—like our sun. As the newer stars formed from the remaining gaseous hydrogen, they often developed what is termed a gaseous protoplanetary disk. After a substantial amount of time, hot gases in this disc cool along with the accretion of dust and heavier material produced by the star and by supernovae to form a protoplanet. At first, a protoplanet may exist in a molten state, but over time, it will cool and develop a solid crust. The solid factor is the crudest of the five fundamental forms of material reality (spacetime, plasma and gaseous elements, electromagnetic radiation, liquid, and solid). In a solid, the pressure of the static mode of the CP has reached its maximum.

In solids, interatomic and intermolecular distances are at a minimum. However, this does not mean that solid matter is as solid as we perceive it with our senses. Essentially the entire mass of an atom is concentrated in the tiny nucleus that holds the heavy subatomic particles —protons and neutrons. The electrons exist in various shells or orbitals outside the nucleus. They are maintained in position by their electrostatic attraction to the protons in the nucleus. However, the mass of an electron is only about one two-thousandths that of a proton and the shells containing the electrons are many thousands of times greater in diameter than that of the nucleus. To put this in perspective, if the nucleus of a typical carbon atom was the size and weight of a lead BB, the electrons would be comparable to dust particles some seventy feet from the BB. Clearly, an atom is almost entirely empty space! The only reason the empty space composing our hand does not immediately pass through the empty space composing a table is that the outermost shells of the atoms in our hand and those in the table are filled with electrons. The electrostatic repulsion between the electrons in our hand and those of the table prevents your hand from passing right through atoms of the table. This creates the illusion of solidity, when in fact common matter is mostly just space.

Nonetheless, a solid object such as a rock or planet is ultimately composed of consciousness—albeit in its crudest form. This is the nadir point of the creation cycle. Consciousness completes the outward or centrifugal phase of the cycle when solid material such as a rock or a planet is formed. This constitutes the inanimate phase of creation and represents the formation of cosmic mind and the macroscopic universe.

The Return Phase of the Cycle

The second or return phase of the cosmic cycle may commence at this point if the conditions on a planet or moon are conducive for life. This phase is the counter or centripetal movement of the creation cycle, in which living organisms form and begin to evolve mind. The same layers of cosmic mind form but in reverse order, and they are contained within the physical boundaries of individual living organisms. The transformation of matter into living organisms takes place gradually due to constant struggle along with the increasing reflection of consciousness.

Life can only develop on a celestial body when there is a proper balance of aerial, luminous, liquid, and solid factors. On a planet like Mercury or Venus, life is not possible since there is an over-abundance of luminous factor resulting in very high temperatures and no liquid water. On the other hand, frozen planets like Uranus and Neptune have insufficient luminous factor reaching their surface from the sun and are too cold for life to evolve. On gas giants like Jupiter and Saturn there is an over-abundance of the aerial factor and essentially no solid or liquid surface for life to evolve. At one time Mars may have been hospitable for the development of living organisms since there is strong evidence that it once had abundant and flowing water and a robust atmosphere. Further exploration of Mars may prove that primitive life forms evolved there billions of years ago. However, today the lack of liquid water and a suitable atmosphere that would shield potential life forms from deadly solar radiation suggest that conditions are not conducive to life.

In our solar system, that leaves Earth where life did evolve. Earth is believed to have formed about 4.5 billion years ago. Initially it was a hot molten body and the heavier elements such as iron and nickel settled into the core of the planet, where they are still found today in a molten state, kept warm by the radioactive decay of unstable heavy elements. The lighter substances gravitated to the surface of the planet and cooled, creating a hard crust. Meteorites and comets containing water and organic compounds including amino acids rained down on the surface of the infant Earth for hundreds of millions of years, creating oceans and continents by about 4.3 billion years ago. A thick atmosphere formed consisting mainly of carbon dioxide, water vapor, sulfur compounds, methane, and nitrogen. The atmosphere helped Earth retain its liquid water and shielded potential life forms from deadly solar radiation. The conditions found on the surface and oceans of the infant Earth were conducive to the formation of life.

According to the spiritual model of creation, life was bound to evolve on Earth because consciousness is inherently creative and as such tries to express itself in the form of unit living beings whenever planetary conditions permit it. However in the absence of an organizing principle (i.e. consciousness/cosmic mind), the incredibly complex combination of atoms and molecules that make up even the simplest life forms would appear to be highly improbable.

The first organisms that formed on Earth might have been proto-life forms that became extinct long ago, but the oldest evidence of life on the Earth is the fossilized remnants of cyanobacteria, dated at 3.5 billion years. These bacteria obtained energy via photosynthesis and therefore utilized energy from the sun. A byproduct of photosynthesis is oxygen, and it is believed that these single-celled organisms were responsible for converting the early atmosphere, which was devoid of oxygen, into one containing oxygen. The oxidizing atmosphere was deadly to many of the other microbes that were present on Earth at the time, but it was a vital component needed for the evolution of more complex organisms, including animal life.

With the first spark of life on a planet, the return phase of creation begins. Only unicellular organisms populated the Earth for a very long time (~3 billion years). However, even single-celled life forms express unit consciousness and possess a rudimentary unit mind. Naturally, they experience constant struggle to survive. In this struggle for survival, the simple unicellular organisms found it advantageous to colonize with other cells and eventually multicellular organisms evolved. The time when multicellular organisms arose on our planet is difficult to determine with any precision, but it was probably about one billion years ago. This was a giant evolutionary step and not surprisingly, it took a very long time before multicellular organisms began to dominate the biosphere.

Organisms evolve in the return phase of creation through the constant struggle for survival. This struggle causes the unit's *objective mind* to become subtler along with greater expression of the *I do* and *I am* qualities of mind. This mental expansion is accompanied by greater physical complexity. On our planet, life forms gradually evolved into increasingly complex species such as plants, invertebrates, vertebrates, and finally mammals.

The return path toward the starting point of the cosmic creation cycle (cosmic consciousness) is not without starts and stops as many life forms come and go when they fail to adapt to changing conditions. There have also been extinction events such as the large meteor strike that occurred 65 million years ago that killed off the dinosaurs. However, the evolutionary movement from simple, less developed life forms into mentally and physically more complex life forms proceeds under the relentless pull of cosmic consciousness.

The Development of Higher Mental Faculties

Less evolved animals are principally guided by instincts, and instinctual behavior is in part governed by cosmic OM of the nonlocal collective mind. Instincts are not learned behavior but result from the architecture of the brain and nervous system and sometimes from an animal's psychic connection to cosmic mind. This would explain why many animals appear to have some degree of cognitive interconnectedness. A new learned behavior is passed along to other animals of the same species via their connection with cosmic mind.[155,156,157]

As organisms evolve, the *I do* faculty of mind develops and they begin exhibiting what could be termed rudimentary intellect. They learn from their experiences and modify their behavior accordingly. For example, a dog learns a series of tricks through training. This learned behavior involves some degree of intellect, since this behavior is not instinctual for the dog.

Higher animals such as apes, dolphins, whales, seals, etc. have a fairly well developed intellect or intelligence. They are able to solve problems using abstract reasoning. That is, they exhibit behavior that is not trial and error but seems to involve the ability to apply previous knowledge to a new situation. Hence, there is a gradual transformation of higher mental functions in animals and not a quantum leap from apes to early hominoids.

An example of animal intelligence was Koko (1971-2018), a female western lowland gorilla that had an active vocabulary of more than 1,000 signs in what her instructor called "Gorilla Sign Language." This allowed her to communicate her emotions and desires with her trainer. Koko gained public attention upon a report of her having adopted a kitten as a pet and creating a name for him.

A border collie named Chaser (2004-2019) was another example of incredible animal intelligence. Chaser learned the names of over one thousand toys. She could fetch a toy by name from a random pile of twenty or more toys. If a new toy was introduced to the pile, she could deduce which was the unknown toy and associate that toy with its name. Chaser had the largest tested memory of any non-human animal.

Evolution of Man

The relentless march of evolution is slow. It has taken nearly a billion years for intelligent life to evolve from the first multicellular organisms. We (Homo sapiens) evolved from our hominid ancestors about 300,000 years ago. But how do we differ from other animals? As we have seen, the unit OM is increasingly transformed into *I do* and *I am* qualities of mind as creatures develop mentally in the return phase of the cosmic cycle. The individual or unit mind grows in proportion to the reflection of cosmic mind on it. As the mind expands, the physical body of human beings has become more complex, with more and subtler glands, in order to adjust to the higher psychic sentiments and demands. The ego, or sense of doership, develops, followed by an increased sense of self-awareness.

Once the sense of *I am* becomes predominant in a creature it is said to be sentient and is fully self-aware and self-determinant. On Earth, we call such creatures human beings.

Since we have a developed sense of self and ego (*I do* quality of mind), we possess free will and can direct our mind according to our desire. Plants and animals, which lack the developed sense of *I am*, cannot act independently. They act according to instincts. Being so guided, lower forms of life do not possess the ability to go against the natural flow of the creation cycle. Humans on the other hand, have the ability to focus their mind in any direction they choose, and mind always takes on the qualities of the object of its attention.

This is because the OM functions to translate the nerve impulses reaching the brain from the gateway organs into externally projected reality. Take for example, the nerve impulses reaching the visual cortex in the brain from the gateway organs of sight (eye, retina, and optic nerve). This objective component of our mind takes on the form of the vibrational patterns of the sensory nerve impulses in the visual cortex, and we have a visual experience. Because of the close connection between mind and brain, any damage to the brain will affect the ability of the OM to perform its function of producing a sensory experience.

And when one experiences thoughts and daydreams, it is the OM that is transformed in the process. Because our mind takes on the qualities of the object of attention, it can be a double-edged sword. We can choose to focus on the subtlety of consciousness and move forward on the cosmic cycle or direct our mind toward the crude material world and move backward toward the unconsciousness of animal existence.

Because the predominant quality of our mind is the *I am*, we are naturally attracted to cosmic consciousness. This is because the individual *I am* quality of mind is merely a reflection of the cosmic *I am*. This cosmic *I am* is the most subtle component of cosmic mind and is nonlocal and permeates the entire universe. An analogy would be a radio station broadcasting music from an antenna on the top of a hill. The radio waves propagate in all directions and are received by the radio in our car. Many other cars are tuned into the station, and although all the cars move around each picks up the same music. Similarly, there is actually only one *I am* broadcast in the cosmos, but our unit minds act like radio receivers and pick up the vibration of this cosmic *I*, and we experience it as our unit *I*.

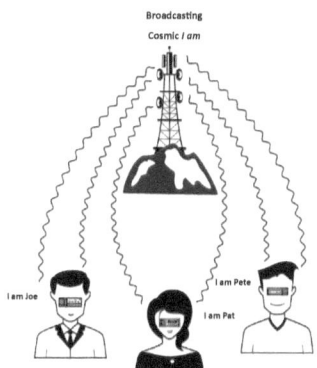

The cosmic *I am* is broadcast universally and similarly to how a radio program is picked up by individual radios, it is received by our unit mind and plays as our personal *I am*.

If we could transcend our limited sense of *I* created by ego, we would experience ourselves as the cosmic *I*. Such an experience is known as self-realization and mystics describe the experience as indescribably blissful accompanied by the feeling that they are one with God.

The Final Journey

The final step in the cosmic cycle is the merger of the unit consciousness with cosmic consciousness. Such merger is a return to the starting point of creation. Merger occurs as the individual mind recognizes more and more that it is an inseparable part of the cosmic *I*. In the process, the individual mind is lost—like a grain of salt dissolving in the ocean. According to the spiritual model of creation, this is the meaning, purpose, and goal of human life—to attain union with cosmic consciousness.

This model for the creation cycle depends on the concept of reincarnation, since the ultimate dissolution of the unit into cosmic consciousness can take more than one lifetime. Our unit consciousness can be considered to have begun billions of years ago in the form of a single-celled organism that gradually evolved in the return phase of the cosmic cycle and eventually attained human status. Then it may take many lifetimes before spiritual union is obtained. In addition to the term self-realization such merger of the unit with the cosmic is also called union with God, enlightenment, liberation, salvation, samadhi, nirvana, moksha, mukti, and satori. According to this model of the creation cycle, this is the ultimate goal of human existence. How this can be obtained in this life is the subject of the final chapter of this book.

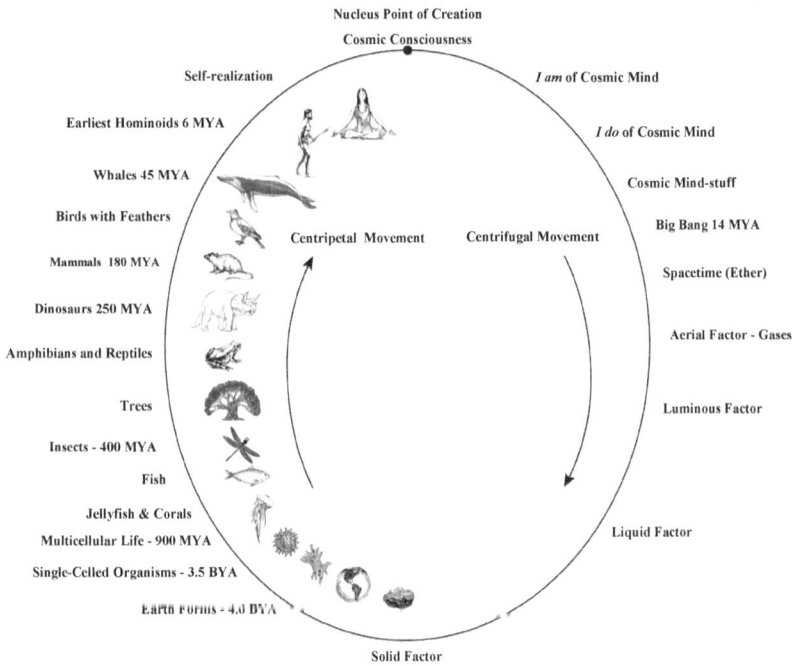

Illustration of the Cosmic Cycle. Cosmic mind (consisting of the cosmic qualities of *I am, I do,* and *objective mind*) is formed before the basic elements of physical reality (spacetime, aerial, luminous, liquid, and solid factor). In the return phase, living organisms develop and evolve eventually leading to sentient beings.

16

HOW SPIRITUAL IDEOLOGY EXPLAINS THE MYSTERIES OF QUANTUM PHYSICS AND RELATIVITY

Cosmic Mind and the Domain of the Wave Function

MOST OF THE PROPONENTS of the materialist worldview now admit that matter is probably not the fundamental building block of reality. Matter is thought to arise from the subtle domain of the wave function. Apparently, matter and the other elements of physical reality (e.g. energy) only emerge from this realm of possibilities when something looks at it. Some physicists in the materialist camp believe that ultimate reality is the wave function of the universe. Since a wave function describes possibilities, such a function would describe all the possible states and interactions that can take place in the universe. This model describes the universe as emerging from a subtle domain of oneness.

As mentioned in Chapter 6, there are no separate parts or arrow of time in this domain. Everything there is entangled or connected—it is a subtle domain of wholeness. Separate parts only emerge following an act of observation in which information is gained about the system. The requirement that someone or "some thing" needs to be informed before this state of oneness is broken implies that mind is involved in the process. In other words, what we call physical reality with its separate parts is an emergent property of this realm and mind is a necessary ingredient in the process.

Interestingly the properties of the domain of the wave function are essentially the same as those of cosmic mind. Below is a table of the properties of each of these domains—one described by quantum mechanics and the other described by spiritual ideology.

Wave Function	Cosmic Mind
The wave function represents the fundamental realm from which physical reality emerges.	Cosmic mind is more fundamental than physical reality and physical reality emerges from it.
It describes the universe as an indivisible whole.	It describes the universe as whole originating from cosmic consciousness.
It is timeless. The past, present, and future are meaningless when discussing this realm. It is only after the wave function collapses by observation that an arrow of time comes into existence.	It is timeless. The past, present, and future are present in the wholeness of cosmic mind. It is only after the first element of physical reality (spacetime) emerges from it that an arrow of time comes into existence.
It is nonlocal. It penetrates and surrounds ordinary reality and is not localized in any part of space but is all-encompassing, everywhere at the same time. An observation is required before a "part" of reality described by the function becomes localized in space.	It is nonlocal. It penetrates and surrounds ordinary reality and is not localized in any part of space but is all-encompassing. It is only after it is converted to spacetime and the other cruder elements of physical reality that separate parts become localized in space.

It is a mathematical representation of the possibilities. It determines the probability that any particular quantum possibility will become "real" (when observed).	It contains all the possibilities that may become reality. However, not all of these possibilities have the same probability of becoming physical reality.
Because the wave function describes only potentialities, some outside agent is required to select which of the possible outcomes will manifest in physical reality. It appears that a nonphysical agent (mind) is required to do this.	It is nonphysical and can be considered the wave function of the universe. It functions as the ultimate observer or outside agent required for the collapse of wave functions and thus for the emergence of the individual parts of physical reality.
The wave function can also be thought of as a field of information. Change in physical reality only takes place when there is a change in how we become "informed" of this subtle domain. In other words, causation can be defined as the movement of information with the wave function representing a field of information with almost limitless possibilities.	Cosmic mind can be thought of as information. The physical state of reality depends on the questions we pose to this higher level of reality. For example, when we apply our intention to do something, we are asking a question of this infinite field of information and in so doing causing physical reality to become manifest in a certain way.
The domain of the wave function is the underlying source of all energy. In a sense, it contains the potential for the expression of almost infinite energy.	Cosmic mind is nonphysical and therefore subtler than matter and energy. It serves as the underlying source of all matter and energy.
The collapse of the wave function does not require energy—just observation or mental intent. Hence, conservation of energy is not an issue when considering how mind could affect matter, but sentient beings might be required to bring reality forth from this hidden domain.	Physical reality is created from cosmic mind. There is no energy transfer in the process. We could say that we live in a conscious universe, and as a result, a physical reality exits without the need for the existence of sentient beings.

All the separate parts that emerge from a wave function are entangled. Physical reality with its associated wave function began with the Big Bang some 13.8 billion years ago; it follows that everything in the universe was then and remains entangled, and that the universe can be described by a wave function.	Cosmic mind is formed from cosmic consciousness before the Big Bang. It represents the organized whole of the universe. Physical reality emerged from it and as a result, everything within physical reality is connected (entangled). Cosmic mind could also be described as the wave function of the universe.

It appears that quantum mechanics reveals a subtle, more fundamental aspect of reality that for thousands of years has been described by spiritual ideology. The parallels between these realms are striking. Both cosmic mind and the wave function can be described as mental realms.[158] But the wave function is also described in mathematical terms as abstract configuration space. As we have seen, the mathematics that describes the wave function contains complex variables that include imaginary numbers. Such numbers don't exist in the real world but are useful for mathematical purposes. This is the reason we call the space defined by the wave function "abstract." Like Einstein's four-dimensional spacetime, we have no way to visualize such a realm with our minds.

Now let's consider the paradox posed by the thought experiment of Schrödinger's cat. The idea of quantum superposition says that the cat must be half-alive and half-dead until we look at it. But what if my wife and I look in the box at the same time? I love cats and she hates them. Which of us decides whether we find the cat alive or dead? Presumably, if the cat is found alive I must have prevailed. This apparent paradox is easily resolved with the understanding that cosmic mind is a singular entity that actually makes the choice. Our individual ego consciousness is part of cosmic mind and our feeling of being a singular being is actually illusory. Cosmic mind transcends individuality.

In this way, the idea of a universal observer easily solves one of the most confounding and illogical implications of quantum physics. If an observer is required before reality can manifest in the material world then *sentient observers are required in order for the universe to exist!* However, it is well known that such observation does not need to occur at any specific moment of time, but can occur retroactively—even many years later. For example, whether a photon behaves like a particle or a wave in a double-slit experiment can depend on whether knowledge of its path was erased or not, and this information may not be gained until many years after the experiment was run.[159] So, if observers are *not* bound by time, why would physicists assume that they are bound by space? Logically if an observation is not "in time," there is no reason to assume it is in space. This implies that an observer who obtains information about a quantum event, which causes wave function collapse, *does not reside in spacetime.* Therefore, the observer is not a person, but must be nonlocal—in a realm more subtle than spacetime—the universal cosmic mind.[160]

This answers the question Albert Einstein asked of Neils Bohr, "Is the Moon there when no one looks?" The answer is, of course the moon is still there if no one looks because cosmic mind is always looking.

Cosmic Mind, Free Will, and Probabilities

The wave function outlines the possibilities that might become physical reality when we are looking at a quantum system. If we square the amplitude of the function, we get the probability that any one of these possibilities will emerge from this domain. But what about cosmic mind? Does it also speak in probabilities? The answer is yes, in a way it does. For example, consider our effort to go to the store to buy a quart of milk. We normally take the most direct route to our favorite grocery store. It's probable that we will choose to go by this route. But suppose that the road is closed due to a water main break. This is an improbable event, but it would force us to go by a different route. Or suppose there has been a flood and the bridge we would go over to the store was out forcing us to take a longer, more circuitous route to the store. Or perhaps we decide to go to another store because of the detour. There are potentially many ways to accomplish our task of getting the milk—many possibilities. But we have some freedom to choose which of these possibilities will manifest as physical reality. There is a much higher probability that our favored route will be taken, but it is within the realm of possibility that we might be forced to take a less likely route.

There are many possibilities in how physical reality may manifest. All possibilities must lie within cosmic mind, but as sentient beings with a modicum of free choice, we have the ability to make choices from the possibilities at our disposal, whereupon all the other possibilities disappear—one of the possible routes to the store becomes reality. However, we are like the scientist that sets up an experiment to observe a quantum system in a certain way. We might predict in advance that there is a 99.999 percent chance of finding the electron inside a box. It is unlikely, yet possible, that it might pop up outside the box. Probabilities, not certainties govern how we experience physical reality—we cannot be sure that the bridge has not collapsed and we need to change our route to the store. In addition, our choices are severely limited. We don't get to choose whether the cat is found dead or alive or whether the moon exists any more than we can choose to fly like a bird to the store.

Clearly, choice is a mental function and results from free will. However, our individual mind is but a reflection of cosmic mind. If we consider cosmic mind to be a higher-dimensional reality, then our individual mind is a lower reflection of it. Thus, our limited experience of the "higher mind" would be like the shadows that play on the wall of Plato's cave. The true reality is cosmic mind within which exist the nonphysical, higher-dimensional forms, ideas, images, and archetypes that give rise to physical reality—*a realm in which probabilities do not apply.*

We experience how our choices seem to affect reality manifesting in our personal universe. We might even state this scientifically and say that as observers or choice-makers, we cause wave function collapse and as a result, physical reality manifests in our little world in a certain way. However, we cannot take credit for making the universe the way it is. We will have to leave this up to cosmic mind. We might participate infinitesimally in how the universe unfolds, but we are *not* a required ingredient despite what John Wheeler proposed in his interpretation of quantum mechanics.

How Spiritual Ideology Relates to Relativity

The special and general theories of relativity (relativity theory) were first proposed by Albert Einstein in the early part of the twentieth century. They describe physical reality as having a number of strange features. One of the most fundamental of these features is that there is a speed limit for light. Light is a small segment of the electromagnetic spectrum but this speed limit applies to all forms of radiation including gravity waves.

In fact, it appears that there is no way for physical forms of information of any kind to travel faster than light speed. Now, you might ask: don't entangled photons pass information between them instantaneously? It is true that the observed polarization state of a photon in our galaxy is instantly communicated to its twin in another galaxy. But since the polarization of the photon in our galaxy could be either vertical or horizontal—and we don't know which until we observe it—this goes for its entangled twin. In general, no useful information can be communicated instantaneously using entanglement.

We might also ask the question of why a creator of the universe would put a speed limit on the physical components of reality including information. Turns out if there were no speed limit for the physical components of reality, then in theory they could reach infinite speed. This would mean that for light, or even matter, they could be everywhere at the same time (nonlocal). Only mind has this capability. The speed limit for light is an insurmountable barrier separating mind and matter that prevents matter from traveling to the past, which could create impossible paradoxes. The hierarchy consciousness » mind » matter is maintained.

For example, imagine getting into a spaceship capable of traveling at "warp" speed—several times the speed of light. According to relativity theory, this is impossible, but suppose we could do it. Now instead of the clock in our spaceship just slowing—(as it does as we near the speed of light)—it would actually reverse, and time would move backward as we accelerate beyond light speed. When we returned to Earth (now in the past) we might encounter our grandmother (now a young girl) and accidently kill her. Then how could we have been born? Many such apparent paradoxes would be expected if people from the future were able to build a time machine that traveled into the past. The fact that we have not been inundated with time travelers from the future wanting to observe our history is also consistent with the idea that time travel into the past is impossible.

While a speed limit for things in the realm of physical reality exists, no such limit applies to mind. Mind is both nonphysical and nonlocal. Our individual mind is a microcosm of the macrocosmic (cosmic) mind. Normally our minds are focused on our brain and behave locally. But under certain circumstances our mind may receive information nonlocally from cosmic mind. There is no speed limit here and no time factor. Just like in the domain of the wave function, everything is entangled or one in this realm. There are no restrictions for the transfer of information if one can tap into the nonlocal cosmic mind.

Not only is there a speed limit for light, but it is constant according to relativity theory. Unlike all other speeds that are additive, light speed is a constant no matter our relative state of motion. This is required if light is to be fully integrated into the spacetime continuum. This means that the four-dimensions of space and time are for the most part not perceptibly different from one another except in the extremes. For example, when a photon moves at the speed of light it moves exclusively through space and not time, but for us moving very slowly compared to light, we seem to move only in time. For things moving between these extremes, they move through space and time at a rate dependent on their relative velocity. The problem for us is that we have great difficulty conceiving of things in four dimensions. Additionally, we are conditioned by our environment to experience the constant flow of time from the past into the future.

We know from our earlier encounter with relativity theory that the constant flow of time is illusory. The fact is that spacetime, when taken as a whole, has events at fixed locations within it that only seem to unfold with time. As mentioned in Chapter 7, what we experience is conscious time—not the time of spacetime—which is unchanging. Because we possess conscious awareness our experience of time is caused by the constantly changing sequence of events we experience. Absent an observer, there would be nothing to witness events, and thus no flow of time.

Einstein's spacetime is just the most subtle element of physical reality. It is created from cosmic mind by the action of the static creative principle on the cosmic *objective mind*. All events for all time exist in the wholeness of spacetime. This is what the ancient sages called the "Akashic Record." In Sanskrit, *akasha* means ether or space. Today scientists use the term spacetime for this subtle element of physical reality. Although cosmic mind is the witness of the entirety of spacetime, knowledge of this sphere of reality is not something the human mind is normally privy to. However, it is possible that by connecting with cosmic mind, we can gain glimpses of events that have or will happen.[161] This could be the mechanism by which some people are able to predict or have experiences foretelling future events. Neither time nor distance would attenuate the transfer of such information—something that is consistently observed in psi experiments of precognition.

Summary

The philosophical implications of quantum physics are deeply troublesome for the materialist worldview. Because of its reductionist approach to understanding reality, it has had to modify its stance that the most fundamental building block of physical reality is matter. Materialists now recognize that there is an underlying realm from which physical reality springs that can be dubbed the domain of the wave function. This domain is characterized by wholeness in which there are no separate parts—just oneness. And this domain is not in spacetime but is a subtler level of reality from which spacetime manifests. This modified model of reality is essentially the same as the one pictured by spirituality. Has materialism finally come full circle and embraced spirituality? Of course the answer is no, because spirituality implies that there is a creative or organizing principle that lies behind reality (consciousness) and such a concept is anathema to materialism. Yet the new materialist model of reality has to cope with the idea of oneness or unity, and it falls well short. It does not explain how everything can be reduced to one—including mind and consciousness, and still manifest as separate parts without anything observing it. This is an unresolved dilemma for materialism—the quantum realm consists only of potentialities unless something looks at it. *Absent an observer, there is no physical reality.* This is where spirituality has the upper hand.

One cannot understate the importance of the wave function to any understanding of quantum mechanics. Yet, the domain of this all-important function is for all practical purposes identical with the domain of cosmic mind. It is obvious that physics has now arrived at a place where it can no longer deny the oneness that is a hallmark of spiritual ideology.

In contrast, relativity theory does not pose a major philosophical threat to materialism because it describes physical reality—not a subtler domain. However, when the totality of spacetime is considered, it also points to oneness since all events, both past and present, are there in the gestalt of spacetime. By proving that space and time are actually inseparable in the oneness of spacetime, it also paints a picture of wholeness for this subtlest element of physical reality. And this oneness applies to matter and energy, which are now known to be essentially different forms of the same thing.

In addition, the reality described by relativity and thermodynamics says that we perceive an arrow of time because entropy is continually increasing. When the clock is turned backwards, we eventually come to time zero—the Big Bang—where entropy (disorder) is at a minimum. We could assign entropy a value of zero at time zero. What this means is that physical reality is maximally ordered at the beginning point of creation. This is exactly what we expect if the universe begins not from a chaotic "bang" but from a state of orderliness. It would be fair to label this ordered state cosmic consciousness, the Creator, God, etc. Why would the universe begin from a maximally ordered state if it arose by mere random chance? Logic says that maximum order is not expected to arise by a chance occurrence. Similarly, it defies logic that something could arise out of nothing.

Both quantum mechanics and relativity theory point inexorably to the fact that oneness underlies physical reality. This is probably the most fundamental postulate of spirituality. If scientists today don't see this it is because they are blind to the implications of both of these great theories.

17

THE MIND-BODY CONNECTION

MANY OF THE PROBLEMS with reducing mind to brain were covered in Chapter 10. The reductionist ontology of materialism requires that every attribute of what we call mind results from brain activity. This assumes that any change in the body or its physiological condition caused by a mental state arises from the brain and its close connection to the body. As noted before neurologists call this mind-body unity. Spiritual ideology provides a sound model for how this unity might work. It starts with a theory proposed by yogic sages dating back many centuries about how the mind is fundamentally constructed.

Psychic Body

Yogis discovered that the human mind has several layers and have categorized them in detail.[162] The physical body can be considered the outermost or crudest layer of our being and many of its functions take place independent of our conscious thoughts.[163] For example, we do not have to think about breathing, heart beating, or releasing digestive enzymes from the pancreas. Most of our glands function autonomically. According to neuroscience, these unconscious bodily activities are governed by a primitive part of the brain and nervous system known as the autonomic nervous system.

Yogic philosophy goes a step further and says that we have a mental body that is intimately tied to the physical body, and that it is the actual controller of most of these bodily functions. As mentioned earlier, this is termed the psychic body. Although other names given to this nonphysical body are mental body and vital body, it should not be confused with the theory of vitalism. This was the idea of how life was infused into dead matter that was popular before advances in molecular biology offered a more scientific explanation for how life processes could be explained using established principles of physics and chemistry. Instead, the psychic body functions as the intermediary between mind and body and is the organizing principle for the proper functioning of the body as a whole.

Evidence for the existence of this mental body can be seen in every living organism, even the most primitive. For example, a slipper-shaped *paramecium* (a single-celled protozoan) can be observed in a microscope to swim about swiftly searching for food using its synchronized oar-like cilia. If it bumps into an object, it recoils and darts off in another direction. Similarly, *euglenas* (unicellular protist) have a red eyespot, a primitive organelle that is sensitive to light, and they are able to adjust their position in order to produce more food via photosynthesis.

A common slime mold (*Physarum polycephalum*) has been found to be able to react and adapt to its environment, and essentially "learn" from its experiences. It also retains a "memory" of those experiences. Yet, it is a single-celled organism whose cognition is obviously not brain based. In *Physarum* and more evolved multicellular organisms including plants that do not possess a nervous system, we see complex behaviors and mental capabilities including the ability to get through a labyrinth, create optimized networks, escape from a trap, and balance their own diet.[164,165] This has caused scientists to rethink whether such primitive organisms possess intelligence. And since they seem to behave in an intelligent manner the question is where such intelligence comes from.

Currently scientists hypothesize that such behavior in primitive organisms results from bioelectricity—cellular currents cause by charged atoms (ions). However, it is reasonable to assume that these patterns of electrical currents are only the physical manifestation of a subtle, highly organized mental body.

According to the model of spirituality, the unifying principle that allows organisms to function and survive is their psychic body. The psychic body functions as the organizing principle as well as an intermediate between their physical body and unit mind. Advanced animals have extremely complicated physical bodies with brains, neurons, sense organs, autonomic systems, and numerous glands that function in an organized and coordinated way. The existence of the psychic body explains how a diverse collection of trillions of individual cells can work together to form a viable organism such as a bird or a dog.

Its existence also explains how different structures and systems develop in an embryo beginning with a single fertilized ovum. All the cells have the exact same DNA, but somehow individual cells know exactly when and how to differentiate into cells that are to become bone, nerve, muscle, eye, etc. There is exquisite control over how cells develop in an embryo. In addition, each cell in the developing embryo needs to know where they are relative to other cells around them so that they can develop into an organ such as a liver or a hand. In other words, they need to have "positional awareness." Absent the "organizing field" of the psychic body, biologists lack a good theory on how the incredibly fine-tuned development and other complex activities required for life and embryo development occur.

This is the argument posed by Rupert Sheldrake in his groundbreaking book, *A New Science of Life* that he published in 1981.[166] He argued persuasively that there exists a nonlocal, nonphysical "morphogenic field" that is responsible for the form, functions, and organization of the trillions of individual cells that constitute a complex living organism, without which it could not function or develop from embryo to adult. The morphogenic field is merely his name for the psychic body.

The psychic body of a living organism is in intimate contact with its physical body. It could be called "psychophysical parallelism." For the brain, with its branching network of intersecting nerves (plexuses), and the physical organs, there are corresponding psychic centers. In vertebrates, the most important controlling nerve plexuses lie along the spinal cord and in the brain. Yogis have identified seven important psychic centers in human beings, which they call chakras.[167]

According to this model, the psychic body serves as the bridge between our brain and our unit mind. When a sense organ in our brain is activated by a stimulus such as a loud noise, our psychic body is vibrated sympathetically by the neurons in our brain and this vibration in turn is conferred to our *objective mind*, and we witness the event. Since all our experiences originate in the brain, the memories of events are also conferred to our mind by the psychic body. Hence, the memory of the event will actually lie in the mind and not the brain. But damage to the brain can interrupt this transfer. Thus, it is not surprising that neuroscientists have been misled into thinking that the brain does it all.

Regeneration is ubiquitous in biology, allowing organisms to repair and maintain the integrity of their bodies. For example, some amphibians can regenerate limbs; other organisms regenerate things like tails, jaws, and retinas. The process is believed to use genetic mechanisms that deploy patterning similar to that of embryonic development. According to the spiritual model, the pattern lies in the psychic body not in the DNA or the proteins coded by DNA. The psychic body is responsible for form, it is the blueprint for constructing a complex organism from a single cell and responsible for the functioning of all its separate parts as an integrated whole.

The healing systems of acupuncture, Ayurveda, and homeopathy depend on the concept of a mental body. These are efficacious healing techniques and should further increase our confidence for the existence of this important and basic layer of mind.

The Other Layers

According to yogis, the human mind has additional layers—each becoming subtler with greater capacity for higher mental functions. Hence, in the next layer of mind, consciousness is more active and it is known as the conscious mind.[168] This layer of mind is active when we are awake. It is home to our desires, fears, and pleasures. It is the crude mind because it is involved with sensing and acting in the external world of physical reality. The conscious mind governs the sense and motor organs, and it deals with bodily needs such as eating, drinking, sleeping, and procreating. It experiences fear and is involved with self-preservation. Man shares with animals the basic needs and qualities controlled by the conscious mind. The only difference is that animals lack the same degree of self-awareness that is enjoyed by humans.

The next layer of mind is the subconscious mind (sometimes called the preconscious mind).[169] This mind is active in both the waking state and during sleep. Dreams are a function of the subconscious mind. Most memories lie in this layer of mind. When activated, memories are expressed in the conscious layer. The current model for how memories are stored in the brain is the network model, but according to spiritual ideology, this is not the full story.[170] The human mind has the same basic qualities that we find in cosmic mind: *I am*, *I do*, and *objective mind*. When we have a sensory experience our unit mind takes on the qualities of the object or thing that the sense organ in the brain experiences. In other words, when we experience the sight and smell of a rose our visual cortex and olfactory centers in the brain receive nerve signals from the gateway organs of the eyes and nose. With the help of our psychic body, the objective component of our mind takes on the properties of the nerve signals to create our subjective experience of the rose.

It is wrong to think that memory is solely an attribute of the brain. Memories are first created in the mind and remain as potential reactions. They are then expressed only when the imaginative power of the mind works through the medium of the cerebral nerve cells. No matter how great the power or potential of the mind may be, it cannot do anything at all without the help of the brain. The brain, psychic body, and mind are intimately involved with all experiences. Damage to the brain can interfere with the process. This means brain damage can result in memories not being created and/or not being capable of expression. Because of the close connection between mind and brain, electrical stimulation of the brain can sometimes bring out detailed memories. And since memories are also stored non-cerebrally, memories of a previous life may occasionally come to the surface on their own or under hypnosis. Such memories normally lie deeply hidden in most individuals.

The subconscious layer of mind is also responsible for higher thinking, contemplation, reasoning, pleasure, and pain. Philosophies, scientific theories, and all sorts of problem-solving activity take place in the subconscious mind. This layer of mind continues to be active during sleep. Sometimes we can go to bed with a problem and wake up having found the solution. Many intellectual and scientific discoveries have occurred when the conscious mind was in a quiet state, allowing for clear awareness of the subconscious mind.

The subtlest layer of unit mind is the unconscious mind. It reflects cosmic mind and it is therefore the transcendent, all-knowing, timeless, collective mind. There is essentially no difference between this universal mind associated with the microcosm and cosmic mind—the mind of the macrocosm.[171] It has none of the boundaries of the individual mind such as time, place, or person. The unconscious mind is responsible for all the higher sentiments and knowledge. The term "unconscious" refers to the fact that most people are unaware of the functioning of this one mind, which is all-knowing or omniscient.

Swiss psychologist, Carl Jung, recognized that humans could tap into a higher mind that is populated by instincts, as well as by archetypes. He called it the "collective unconscious." His evidence for the existence of this collective mind linking all humankind was based on the observation that among cultures with no historical influence upon one another, the basic symbols and myths were universal. Jung called the predisposed patterns for particular myths and symbols "archetypes." He argued that they exist in the collective unconscious of humankind and are similar to animal instincts in that they influence the basic structure and organization of the human psyche. Jung identified several archetypes, such as the hero, the mother, the child, the devil, and the trickster. He also offered numerous examples of myths and symbols that arose in cultures with no direct link to one another.[172]

Joseph Campbell was another influential thinker who believed in the psychic unity of humankind and its poetic expression through mythologies. He provided numerous examples of universal symbols and myths in his book, *The Hero with a Thousand Faces*.[173] Like Jung, he postulated that there had to be an all-pervasive collective mind to explain these facts. The existence of cosmic mind provides a simple explanation for why these archetypal patterns and myths exist.

The unconscious mind is the seat for all the higher mental functions. These include magnanimity, humility, serenity, gentleness, mercy, intuition, discrimination, non-attachment, conscience, creative insight, and spiritual ecstasy.

Experience of the limitless unconscious mind gives expanded awareness, knowledge, and energy. It is rare for an individual to penetrate deeper than their subconscious mind, but if they do, it is a profound and moving experience. For a brief moment, they may experience the uninhibited flow of ecstasy, supermundane knowledge, and unconditional love present in the unconscious mind. This may last but a short time because the unrelenting distractions of the physical world inevitably draw them back to mundane awareness. However, it is no surprise that such a mystical experience will forever change the way they view reality.

Finally, yogic philosophy describes a self or *atman* that lies beyond the unconscious mind. It is the witnessing entity of our individual being. This immortal self is created by the reflection of the cosmic *I am* on our mental plate. The *atman* can also be termed the self, unit consciousness, or soul.

Some Implications of the Model

Earlier the question was asked of how a mental state suggested by a hypnotist such as the expectation that the palm of one's hand was being burned by a candle could actually cause the skin on the palm to develop a blister. How this could occur from a purely physicalist perspective would have to include some mechanism by which nerve signals from the brain cause skin cells to undergo changes as though they were actually damaged by excessive heat. No such mechanism is known. However, if body is actually just the outermost shell of mind, we might postulate that our conscious or subconscious minds could under certain circumstances effect a change in the physical layer of mind (the body). This would be a mind-mind effect instead of a brain-body effect.

A mechanism for how this might happen depends on the hypnotized person obtaining temporary concentration of their mind, and it taking on the form of the blister they imagine would form. In such a state, the physical body could be affected in such a way that it undergoes changes consistent with the mental picture produced by their subconscious mind. This is because each layer of mind has a corresponding or sympathetic vibration with neighboring layers and changes in the psychic body directly affect the physical body and vice versa. Just like two tuning forks that undergo harmonic resonance when one fork responds to vibrations from its neighbor, so too do layers of our mind resonate in harmony with one another. For example, a disease of the body such as pneumonia will also manifest in the psychic body and will affect the conscious and subconscious layers of mind. Treatment for such a condition could be purely physical—such as taking an antibiotic. Alternatively, treatment might include a mental component such as taking a placebo, assurance from a doctor that our condition is nothing to worry about, or our mental expectation that our prayers for a quick recovery will be answered. Because of the close connection between the various layers of mind, a treatment that affects the psychic body such as acupuncture might be effective. As a result, purely mental treatments can be efficacious for curing diseases. For example, Christian Scientists believe that disease is really a mental error rather than a physical disorder and many practitioners of Christian Science have good success treating diseases accordingly.

The unity of mind and body explains how mental states can affect the body. For example, it is well known that psychological feelings such as hopelessness and depression can indirectly damage the body while feelings of elation and laughter can improve health. The expectation that a medicine will improve health has this exact effect. This is why the placebo effect is so powerful, and why it can produce physiological changes due to expectations alone. The other mind-body effects mentioned include psychosomatic illness, stigmata, skin writing, hypnotic effects, and how patients with dissociative identity disorder (multiple personalities) may exhibit different physiological conditions. There is a myriad of such effects of this type in which some condition of the mind affects the body. The alternate hypothesis of materialism is to say that mere expectation— that does not have any known physical foundation—nonetheless causes a physical change.

Mind as a nonphysical component of being with a close connection to consciousness also solves the problem of the unity of sensory experience. Since there is no anatomical or brain basis that adequately explains how sensory inputs are unified into a coherent experience, this is one of the functions of mind. Mind is unitary unlike brain with its many parts that would need to work together to provide a unified experience of physical reality.

Our brain has evolved over the millennia. First to evolve was the primitive brain or brain stem that controls autonomic responses and conveys motor and sensory pathways from the brain to the body and from the body back to the brain. Next to evolve was the cerebellum (little brain), followed by the cerebrum, and finally the cerebral cortex or outermost layer of the brain. As the brain evolved, the layers were built one over the other like layers of a pearl. In addition, within the brain are many different structures all having different functions and anatomical features. Neuroscientists postulate that the whole thing comes together in producing a unified conscious experience because of a hierarchical arrangement of brain networks. A simpler and logical alternative to this model is that the nonphysical component of our being—mind—serves this function while at the same time supplying the conscious awareness that accompanies our experience of physical reality.

The spiritual model solves the problems inherent in the materialist model of how the brain generates conscious awareness, intentionality, and subjective experiences (qualia). The hard problem of neuroscience of how the brain creates conscious awareness is easily solved by this model. Not only are we conscious beings but our experience of physical reality is accompanied by a full spectrum of feelings and emotions. The materialist model suffers from an inability to explain how a three-pound moist computer can produce subjective experiences of love, hate, delight, disgust, awe, and bliss as well as a sense of self-awareness. But this is easily explained by a model that identifies mind and not brain as the ultimate experiencer.

Aristotle recognized that cognitive awareness and volition go together like two sides of a coin. Volition or intentionality depends on the sense of there being an *I* that wants to do something. According to spiritual ideology, the sense of *I am* is the first and most basic component of mind and is created by the sentient creative principle as it operates on consciousness. The other hard problem for materialism is how you could create intentionality by physical means. For example, how could you program the feeling of "I-ness" into a computer so that it could display intentionality?

Lastly, the spiritual model for mind provides a logical way to understand the self. Earlier we considered the fact that unlike consciousness that may flicker on and off, the thing that remains constant in our life is the feeling of being the same person today as we were yesterday—or for that matter since the day we were born. That is, we possess self-awareness. Why is it we feel we are the same person despite the fact that physically we change constantly? It is not because there is something about the network or pattern of our brain that does not change. In fact, there is almost nothing the same about our brain now as compared to when we were a child of three. Both the neurons and their connections are replenished on a regular basis.

Perhaps selfhood depends on our memories that tie things together into a coherent picture. This is also unlikely since even people with amnesia, who might not remember who they are, still recognize that they are a unitary human being with thoughts and desires, and they know they have existed since their birth—it's just that they cannot remember their identity. Finally, why does a sense of self exist in out-of-body and near-death experiences when the brain is shut down?

The physicalist approach to neuroscience fails to explain why we feel we are the same person throughout our lifetime when there is essentially no physical aspect of our being that remains constant. The more logical approach of spirituality explains that our self-awareness results from our experience of a personal *I am*—a reflection of the cosmic *I am* on our mental plate. If we could just experience this *I am* in its pure, unqualified state free of ego, then it would be experienced as the higher self or *I am* of cosmic mind—since in reality there is only the cosmic self.[174] The personal self is illusory—a product of identifying with the *I do* and *I have done* qualities of our unit mind (ego) and its focus on our body. The egoless experience of reality—self-realization—is unitary, nonlocal, timeless, and indescribably blissful.

18

UNDERSTANDING THE PARANORMAL

THE DEFINITION OF PARANORMAL is phenomena that are unnatural, psychic, or are unexplained by science. Such phenomena include out-of-body experiences (OBEs) and near-death experiences (NDEs). In addition, mystical experiences, psi phenomena, and reincarnation memories fall into this category. For the most part such phenomena are incompatible with the materialist worldview, and the advocates of materialist ontology just dismiss the veracity of all such phenomena. In Chapter 23, we will explore some of the reasons for this mindset. However, there is such overwhelming scientific evidence for the existence of psi that we can safely assume that psi is real and that people's experience of the paranormal is not some form of mass delusion but deserves an explanation.

Theory to Explain How Mind can Function Free of the Body

Experiences in which the mind appears to continue to function after leaving the body, or at least when the brain is shut down include OBEs and NDEs. It is logical that mind could continue to function under such circumstances if it is a nonphysical aspect of our being and not just a byproduct of brain activity. However, an explanatory mechanism for how the mind could continue to acquire information about events occurring after it is no longer receiving information from the sense organs faces several challenges. Foremost is the fact that our unit mind is normally focused on our brain. Our experience of physical reality comes to us through our brain via our sense organs. In order for us to have lucid experiences while the brain is shut down, there would have to be a different way for us to experience physical reality.

Let's assume that the brain shuts down due to a lack of blood flow following cardiac arrest. Yogis claim that this can temporarily cause the psychic body to sever its close parallelism with the brain. Under these unusual circumstances, the psychic body along with the other higher layers of our mind could still receive the inferential waves produced by events taking place nearby—such as attempts at resuscitation. Normally the psychic body is the intermediary between nerve impulses in the brain and our mind. However, as mentioned earlier, yogis claim that it is possible for the psychic body to leave and travel away from the physical body as long as it is attached to the physical body by what they term a subtle psychic cord—i.e. astral projection. If this cord is severed then it will result in death. This is the likely mechanism by which lucid awareness of physical events during an OBE could be experienced by a person who might be considered "clinically dead."

Obviously, NDEs include additional features such as entering a tunnel with a brilliant light at the other end; feelings of being in the presence of a Supreme Being with unconditional love; complete freedom, bliss; and a life review, etc. These experiences undoubtedly could occur as the mind progresses the path toward final separation with the body. However, once the subtle connection between the psychic and physical bodies is severed then death will occur with no possibility of consciousness returning to the body.

In a mystical experience, there is normally no close brush with death but again a separation of the unit mind from its preoccupation with brain. There is a transcendental experience of unity as the mind experiences its own *I am* component as its cosmic counterpart. Since our personal sense of *I am* is but a reflection of the cosmic *I am*, the experience is transcendental and unitary rather than personal. Here the unit mind is freed from the incessant "chatter" of the conscious mind that is generated by the brain, and the dream-like experiences of the subconscious mind, and freed to experience the oneness of the unconscious/cosmic mind.

Theory to Explain Psi Phenomena

From the scientific evidence concerning psi phenomena, we know that psi information must be conveyed nonlocally. This means that the information is not diminished by distance, and that time is not a factor in its transmission. The latter requirement means that the information about an event could be gained before or after the event takes place. Like the domain of the wave function/cosmic mind, psi phenomena appear to follow the rule of nonlocality and timelessness.

Unlike the elements of physical reality in which the transfer of information is restricted to the speed of light, the nonlocality inherent in cosmic mind allows for the transfer of information with no restrictions concerning time, place, or person. Since everyone shares the same unconscious mind—which is the same as saying that we have a connection with cosmic mind—the mental impressions characterized by telepathy, remote viewing, and precognition can be understood to be information obtained from this universal or one mind. Thus, we have a rational explanation for these psi phenomena. In fact, these phenomena could easily be the same phenomenon—just slightly different ways of characterizing information gained via the unconscious mind. Yogic sages have identified this as the means by which information may be obtained nonlocally. The term they use is "supramental vision." Instead of experiencing physical reality through the sense organs, it is a way of experiencing aspects of reality directly with the mind.

A theory for explaining psychokinesis depends on explaining how mind, which is nonphysical can affect physical reality. As mentioned earlier, the best theory for this goes back to how mental intention can collapse a wave function causing physical reality to manifest in a specific way. This is the likely mechanism for how the nonphysical mind interacts with the physical brain. For example, consider a quantum system that we do not look at. It will go on happily behaving like a wave spreading itself throughout the entire universe. The moment we choose to look, the quantum instantly reverts to its particle nature and pops into physical reality. As we have seen, the most popular interpretations of quantum mechanics say that mind or consciousness is required before this can occur and things manifest in physical reality.

Psychokinesis is defined as mind affecting matter, and it could work similarly. For example, we apply our intention to get more zeros than ones from a random number generator (RNG). This has the effect of altering the radioactive decay of the unstable isotope used in the RNG. Radioactive decay is a quantum phenomenon and how it manifests in physical reality could be influenced by mind. Because of our intention, the normal randomness of the RNG might be skewed toward more zeros than ones, or vice versa for a short time. Since mind is nonlocal, it would not matter when our intention was applied or how far we were from the RNG—both of which are observed in experiments. Mental intention could affect a macroscopic object like a die similarly, but its effect might be weaker, which is precisely what is observed in experiments.

Tapping into Cosmic Mind

Since cosmic mind is the all-knowing repository of all knowledge both of the past and the future, it is logical to assume that some of the more notorious prophets and seers of the past had at least a limited ability to tap into its untold treasures. The Bible is rich with such personalities and other examples include the Oracle of Delphi and Nostradamus. However, a more contemporary personality whose incredible abilities to diagnose people's medical condition and make accurate predictions of future events was Edgar Cayce. His readings of people's medical condition and cures were extensively studied.

Cayce was born in Kentucky in 1877 and received only an eighth-grade education before he was called to work on the family farm. He came from a devoted Christian family and was a devoted Christian himself. He would read the Bible cover to cover every year and regularly taught Sunday Bible School. It is therefore ironic that the Christian Church eventually labeled him as a false prophet and his work an inspiration of the devil.

At the age of twenty-three, after suffering from chronic laryngitis, Cayce sought help for his condition from a hypnotist. He was told to perform self-hypnosis in order to obtain a permanent cure. After entering a self-induced trance, he described in detail the ailment and its cure. This was his first psychic reading, and over fourteen thousand were to follow before his death in 1945. He performed his readings while in a trance and after awakening had no recollection of what he said. Most of his readings were for people seeking medical diagnoses and treatments. All he required was the individual's name and location. He would then give an accurate description of the illness or physical problem as though he had X-ray vision and then offer a treatment for the condition. The accuracy of his readings was astounding and the terminology he used was not that of an unschooled individual but of someone highly trained in medicine.

Cayce also did readings on geology, chemistry, physics, electricity, history, political science, sociology, and anthropology. Interestingly, he confirmed the validity of reincarnation and indicated that some medical conditions and birth defects could be traced to events occurring in previous incarnations. By the time of his death, he had performed past-life readings for over twenty-five hundred people. He would sometimes trace an individual's existence back thousands of years through numerous incarnations, pointing out how the person's current condition was influenced by past lives. Thus, Cayce strongly confirmed the law of karma and the reality of reincarnation.[175] This was probably the principle reason that he was eventually denounced by authorities of the Christian Church.

Cayce also made numerous prophecies that turned out to be uncannily accurate, such as the discovery of the Dead Sea Scrolls, the stock market crash of 1929, and the start and combatants of World War II.[176]

Cayce's uncanny ability to perform psychic readings could be explained by his ability to tap into cosmic mind while in a trance. In such a state, with his own mind suspended, he could receive the supermundane knowledge of the omniscient cosmic mind and communicate such information to his assistant.

19

LIFE AFTER DEATH

According to spiritual ideology, our purpose in life is to discover and fulfill our deep, innate potential. This is a journey of transformation that is at the mystical heart of all religions. It is a journey to meet the Higher Self and simultaneously meet the Divine. In other words, our ultimate goal in life is to obtain union with cosmic consciousness. Naturally, we find many terms used to both describe the process and the attainment of union, but they all boil down to discovering who we actually are and becoming *one* with the Creator of this universe when we leave our physical body.

The idea that we must eventually complete the creation cycle by attaining union with cosmic consciousness is probably the most fundamental feature of spiritual ideology. Assuming it has taken some 3-4 billion years for our physical and mental structures to have evolved from unicellular ancestors it is not unreasonable to expect that it might take more than one lifetime to develop the mental elevation required to set aside our material and ego goals in favor of the goal of spiritual union. Thus, reincarnation is an important element in spiritual ideology along with its corollary—the law of karma. So in order to understand why there is life after death and what form it takes, it is best to explore in greater detail this simple rule of how our actions echo back upon us.

Law of Karma

Newton's third law describes the law of action and reaction for physical bodies—for every action, there is always an equal and opposite reaction. The law of karma expresses a similar idea in the psychic realm with the caveat that reactions to actions performed may be stored in the mind and experienced later. Simply stated then, the law of karma is "you reap what you sow."

All the world's religions subscribe to the rule of karma even if they do not specifically endorse the doctrine of reincarnation. The idea that "what goes around comes around" is the reason we should follow the Golden Rule, since good actions beget good reactions and bad actions beget bad reactions. People who understand how actions, either good or bad, inevitably return to affect them in this life or in a future life have a reason to act morally. That is, they are aware that there is no escaping the consequences of their actions. Moral behavior is in our own self-interest!

The mechanism for how reactions are stored in the mind involves our *objective mind*. It is vibrated by any mental or physical action, and it retains an impression or reactive momentum for that action (Sanskrit: *samskara*). Such a stored reactive momentum holds the potential for future action. When a stored reactive momentum is expressed, it is said to be "exhausted" or "burned." In other words, the fruit of our action has ripened. Our karma is nothing but the totality of all our reactive momenta. The burning of a reactive momentum returns the mind to a "purer" state. An analogy might be to make a dent in a hollow rubber ball with our finger. The depression may last for a while, but if the ball is massaged with the hand, the dent can pop out, creating a symmetrical ball once again. In this example, the ball represents the mind and the dent represents a reactive momentum. When the dent pops out, it is burned.

Reactive momenta result from our mental and physical actions. They are stored as potential mental energy in the subconscious layer of our mind along with memories of the actions. The expression or burning of a reactive momentum is accompanied by the release of kinetic mental or physical energy equivalent to the mental energy or impression that created it. Although the amount of mental energy may be the same, the type and quality expressed will normally be different. For example, if we do a kind deed for a stranger on the road by helping them change a flat tire, the reaction to such kindness will be stored in our mind and perhaps come back to us later in the form of help someone provides to us. However, it will probably not be help changing a flat tire. In a subtle, often unconscious way, the reaction teaches us that it is good to do good deeds and be nice to other people.

Similarly, a bad deed will ultimately bounce back, creating hardship, pain, unhappiness, or suffering. The mental disturbance reminds us either consciously or unconsciously that the action that created the reactive momentum had bad consequences, and we will be less apt to repeat such an action. This is the basic mechanism behind the law of karma. In other words, this law of action/reaction is a reward/punishment system by which we learn to act better, be wiser, more selfless, and move forward on the path to self-realization and the ultimate attainment of unity with cosmic consciousness.

From a religious perspective, this model of the afterlife makes a lot of sense. The law of karma offers a simple and logical explanation for why some people seem to be born into a life of happiness and ease while others seem destined for suffering. For example, we witness children who are born into a life of hardship and suffering due to no obvious fault of their own, while other children are born with a proverbial "silver spoon" in their mouth. Since we have no knowledge of a child's past lives, we remain mystified by such so-called "accidents of birth." However, the picture of the afterlife we have painted here says that in reality, everything in the universe is connected. There are actually no accidents of birth. The reason some individuals seem to have a mountain of problems to overcome while others seem destined for happiness and success is that they brought it upon themselves by behavior in this or in past lives. The fact that we do not remember our negative deeds from the past does not free us from having to reap their reactions. We are blessed that once we burn a negative reaction it is gone forever, and at a subtle mental level, we are reminded not to repeat the action that brought on the pain. The burning of old reactive momenta and the production of new ones may go on for many lifetimes until we attain union with cosmic consciousness.

We have free will and can direct our mind and act according to our whims. However, such ability to make choices comes with the responsibility that we create both good and bad reactive momenta. Unfortunately, we do not have a choice as to when and how they are burned. The greater the mental vibration or intensity of an action, the greater will be the effect when the reactive momentum created by the action is experienced. Strong desires also create strong reactions. For example, a girl witnesses firefighters saving a child from a burning house and develops a strong desire to become a firefighter herself. After many years and extensive training, she may satisfy her urge to serve her community as a firefighter.

According to the model, actions performed without the *I do* of ego do not create reactive momenta. Therefore, the actions of others, actions performed unconsciously, and acts of God, such as floods or windstorms, do not create reactions in our mind directly. However, such experiences may create other reactions.

Why We Reincarnate

In Chapter 14 we discussed the how and why of reincarnation. According to the model of the afterlife postulated by spiritual ideology, when we die we lose parallelism between our physical body and mind. Unless we attain union with cosmic consciousness at this time, our unit mind continues to exist as a bodiless mind. Lacking a physical body, brain, and a psychic body, this bodiless mind is unable to interact in any way with physical reality, but it still has the three basic elements of mind: *I, I do*, and *objective mind*—with its unburned reactive momenta.[177] Since reactive momenta were created by our interaction with physical reality, they can only be experienced in a physical body. Rebirth into a new physical body is therefore necessary in order for us to be able to experience our unburned reactive momenta following death.

Cosmic mind is responsible for finding a suitable new body with an optimal environment so that we might continue on the path to unification with cosmic consciousness. This could take place almost immediately after death or after some years. Once a suitable home consisting of a fertilized ovum is found, our bodiless mind becomes associated with it and will mature into a baby and be reborn.

Most of the scientific evidence for reincarnation comes from the accurate descriptions of previous lives by children. The question is how this is possible. Since the bodiless mind contains both the reactive momenta and noncerebral memories of past events, it is sometimes possible for a person to recall experiences from a previous life. Fortunately, such memories normally lie deeply hidden in the subconscious layer of our mind. As children, our memories of events that took place before the age of three are sporadic at best. But for a few children—especially if they suffered a traumatic death—memories of living in a previous body may surface. This can be a cause of anxiety, and in the West, the child's parents will usually try to assure the child that these memories have no basis in reality. It is fortuitous that such memories of living in another body begin fading by the time children reach the age of five.

For adults, memories of past lives may come out during dreams, deep meditation, or during hypnosis. It was mentioned earlier that a common clinical treatment for persons suffering from phobias or neurotic fears is hypnotic regression to a time when they first experienced an incident associated with the intense fear. If the traumatic experience that triggered this fear can be relived under hypnosis, it can cause a catharsis that helps to eliminate the phobia whether it occurred in this or a previous body. Except for medical treatment, we should probably avoid trying to remember our past lives. It can create anxiety and can divert our attention from the job at hand, which is to know our self in the here and now.

Summary

Assuming that the purpose of life is to realize who we really are, it follows that the Creator of this universe would include a system whereby actions that reinforce our connection with him reward us with happiness while actions that distance us are punished—the law of karma. In other words, actions that draw us closer to self-realization (e.g. selfless service, spiritual practices, and meditation) produce the greatest happiness while actions that move us farther away from self-realization produce negative reactive momenta and result in pain and suffering. However, both our good and bad reactions create karma. This is why Buddha called good reactions gold chains and bad actions iron chains.[178] He said that the only way to escape the cycle of continuous death and rebirth is to "awaken" to the realization of who we really are.

The fact that many people have veridical experiences of living in another body demands an explanation, not just the blanket denial of materialists that life after death is impossible. The body of evidence for reincarnation is both extensive and scientifically sound. The totality of the evidence is overwhelming and is bolstered by the evidence from OBEs and NDEs that consciousness can continue when the brain is shut down.

Spiritual ideology has an explanation for how life continues following death, and it is reincarnation. Reincarnation is both reasonable and is a fundamental tenet of not only spiritual ideology but all the Eastern religions. This ideology postulates that we must eventually complete the creation cycle and attain the indescribable ecstasy and everlasting liberation of merging with limitless cosmic consciousness—and this can only be accomplished in a human frame.

20

EVOLUTION AND THE MYSTERY OF HOW LIFE BEGAN

How Life Began

One of the greatest mysteries facing science today is how living organisms originated from nonliving matter. The process is known as abiogenesis. Scientists have proposed several hypotheses of how abiogenesis could have occurred on Earth some 3.5-4.0 billion years ago. One theory that is popular among scientists is that complex molecules capable of replication came first. Another theory proposes that molecules capable of metabolism came first. All the theories begin with the well-established fact that the Earth at that time had abundant organic compounds and energy sources from both the sun and hydrothermal vents that could cause the needed chemical transformations. They also agree that nonliving chemical compounds must have undergone an evolutionary-like process that selected such compounds for increased complexity and those compounds had the ability to pass their increased complexity on—i.e. be replicable. They also agree that proto-life forms were probably RNA-based instead of DNA-based because RNA can replicate, assemble proteins, and catalyze some of the chemical reactions required for replication and metabolism of living organisms. Such proto-life forms would then have become extinct long ago.

Of course, theories about the origin of life are fine and scientists feel that it is only a matter of time before the whole story is understood. But many of the mechanisms and chemical transformations involving nonliving molecules required for life such as molecular self-replication, self-assembly, autocatalysis, and emergence of cell membranes are poorly understood. That atoms and molecules can self-organize is a well-known fact. For example, soap bubbles form spontaneously as detergent molecules assemble themselves into spherical bubbles because this is a more stable structure; and sodium and chloride ions assemble themselves into a symmetrical cubic crystal lattice. Similarly, a cell membrane is composed of a lipid bilayer that is similar to a soap bubble and lipid bilayers can form spontaneously in water. The formation of a cell membrane that encapsulates and protects the incredible machinery of life is a necessary component of the abiogenesis story. Furthermore, it is not surprising that the liquid that fills all cells (cytoplasm) is very similar to the composition of the primordial ocean that is thought to have existed four billion year ago.[179]

However, for abiogenesis to occur scientists believe that not only did molecules self-organize but that they *evolved* into molecules that are more complex. The problem is that this has never been observed in the laboratory. RNA can do many amazing things such as replicate itself; catalyze chemical reactions both for self-assembly and for protein synthesis; and undergo mutations that increase its effectiveness at doing these things. But how this extremely complex molecule could form in the first place from free nucleotides swimming inside a bit of sea water encapsulated within a lipid bilayer is a mystery. If this problem could be solved, it would not explain why the RNA started to utilize the free energy available in the environment to perform the myriad metabolic functions that are also required for life. Life is incredibly complex and to explain its origin as a chemical game of chance is very challenging.

This can be seen by considering the probabilities that some of the required ingredients for life came together by chance. There is a very small probability that proteins would arise by themselves. This goes for nucleotides, which are the building blocks for nucleic acids such as RNA and DNA, coming together to form proto-RNA. Since both would be needed in a proto-life form, we need to multiply the probabilities that these two low-probability events arise simultaneously. The result would be a probability very close to zero. At the same time, we would need a cell membrane to appear. Then proteins having specific functions such as catalyzing metabolic reactions and for the linking of the nucleotides that constitute the backbone of RNA are needed. We arrive at a probability that is essentially zero. The problem hypothesizing that such a low probability event might occur is not insurmountable, but the mechanism would also involve a circular scenario—RNA is needed to synthesize proteins and proteins are needed to synthesize RNA.

We are left with the need to hypothesize that the origin of life out of complete randomness requires a degree of fine-tuning similar to that required to make a universe that would be conducive to the formation and evolution of sentient beings in the first place. In addition, computer simulations show that the long-term habitability of a planet needed for the evolution of sentient life—roughly three billion years—requires both luck and stabilizing feedbacks. Something that only occurred in just one out of 100,000 earth-like planets in the simulation.[180] Thus we would need to expand the anthropic principle to say: (1) we live in one of the extremely rare universes that is finely tuned for the possible existence of life; (2) against enormous odds our planet was one of the rare places in the universe that had suitable conditions for the development of life; (3) life did develop here; and (4) against enormous odds our planet remained hospitable for over three billion years. Considering the odds against these unlikely events occurring, we are most fortunate indeed to be alive!

On the other hand, the spiritual explanation for abiogenesis is very simple and straightforward. Consciousness is transformed into cosmic mind. Cosmic mind is transformed into matter. Given suitable conditions like those found on Earth some 3.5 billion years ago, consciousness, which is inherently creative in nature, tries to express itself in the form of living organisms. In other words, the potential for consciousness, mind, and life lie dormant in matter—and under suitable environmental conditions, life will manifest.

Mysteries of Evolution

The modern theory of evolution is called neo-Darwinism. It is a synthesis of the theory of evolution proposed by Charles Darwin along with the new understanding of genetics that includes the role of DNA. Neo-Darwinism assumes that there is no need to hypothesize an "organizing principle" at work behind the evolution of species. This is not to say that

Darwin's theory of evolution and the existence of a Creator are mutually exclusive, but that a Creator might not be required. Moreover, if such a Creator did exist, why was his work so obviously one of trial and error and having so many mistakes?

Neo-Darwinism proposes that life arose from inanimate chemicals, and the evolution of organisms into complex and various species, can be reduced to known physical and chemical processes that are passed on genetically to those organisms that best adapt to their environment. The reductionist doctrine of evolution is now accepted by nearly all biologists, and one of its best spokespersons is Richard Dawkins who wrote the book, *The Blind Watchmaker: Why the Evidence of Evolution Reveals a Universe without Design*.

There is no doubt that neo-Darwinism outlines much of the basic workings of evolution. There is overwhelming evidence that all life on this planet arose from a common ancestor and that species evolved from more primitive ancestors. Additional support for the theory comes from embryology. In most animals, the early stages of embryonic development are quite similar. But is neo-Darwinism the full story?

According to the theory, evolution should be slow, continuous, and incremental. There should be a continuous fossil record of all stages of evolution. According to the theory, there should be many thousands of fossils found for intermediates, thus filling up gaps between species. But very few cases of transitional fossils have been found.[181] In fact, there are numerous gaps in the fossil record. This strongly suggests that there has been discontinuity or unexplained "jumps" in the evolution of species. If evolution is mostly continuous but at the same time has discontinuous jumps, it signals that there is biological creativity at work. This is a definite sign of some form of intelligence at work behind the scenes. In other words, neo-Darwinism works fine when explaining evolution as slow and continuous, but it offers no mechanism for how the tempo of evolution can be sped up creating jumps or significant gaps in the fossil record.

One example of this was the explosion of fauna, seemingly abruptly and from nowhere, that occurred during the early Cambrian Period (~500 million years ago). Darwin recognized that this relatively short evolutionary event (20-25 million years) might be one of the main objections that could be made against his theory of evolution by natural selection.

Similarly during the Eocene Epoch of the Cenozoic era, from approximately fifty million to thirty-five million years ago, the so-called "Age of Mammals" occurred. During this relatively short evolutionary period over twenty-four different orders of mammals evolved—everything from primates to whales.

Speciation also implies that there are unexplained jumps in evolution. Darwinian evolution is normally gradual. Yet, the most common way species are differentiated is by their inability to procreate with one another. The gradualism of evolution that is required by the model of random variation and natural selection of beneficial traits fails to explain how macroscopic traits emerge in different species and why they would not be able to interbreed and create hybrids instead of distinct species.

Another problem for neo-Darwinism is how to explain the slow, stepwise evolution of irreducibly complex systems and organs. This is the argument made by biologist Michael Behe in his book *Darwin's Black Box: The Biochemical Challenge to Evolution*.[182] The biochemistry of life is so complex and interwoven that it is incredibly unlikely that all the elements for some complex organs and systems could come together merely by chance mutations when there is no advantage to an organism until a complete and working organ/system is created. Such a complex system cannot be reduced to individual parts or simple stepwise mutations because there is no advantage conferred to the organism until a *complete and functioning system evolves*.

In fact, most mutations turn out to be harmful so it is more likely that genetic variations would be removed from the gene pool. As a result the probability that so many beneficial gene variations could come together to form even a simple organ such as a gall bladder would take far longer than geologic time.[183]

An example of an irreducibly complex system is the blood clotting system in mammals. No less than eleven enzymes are involved in this intricate cascading system with several feedback loops. If any one of the eleven enzymes is missing or defective, it will be a death sentence for the organism. This is seen in hemophiliacs, who have one defective gene resulting in the loss of a single blood-clotting enzyme, and they inevitably die young unless they receive modern treatments. Such an irreducibly complex system requires that essentially all the components are present before there is any advantage conferred to the organism. Biologists struggle to explain how such systems could evolve in a stepwise manner via random mutations of DNA with selection of the fittest. The odds against all the required components of the system arising simultaneously are enormous.

In addition to blood coagulation, other irreducibly complex systems include the bacterial flagella, cilia in eukaryotic cells, the mammalian immune system, and the eye. Motile cilia function requires 200 proteins, and they might all have to be produced at the same time before the cilia function and there is a benefit for the organism. For the mammalian eye, the number of mutations needed is even greater. But biologist like to refute the argument for some form of guided or intelligent design behind evolution by pointing out that the ability to detect light and hence to develop eyes is of such survival benefit that it may have evolved separately dozens of times over the history of life on Earth. In fact, this probably occurred, but it could equally be used as an argument for guided evolution.

There are many other examples of evolutionary marvels that biologists struggle to explain using the strict materialist approach of neo-Darwinism. Just one example is the wasp *Pepsis marginata* (tarantula hawk) that can only nourish its young from a single egg on a tarantula that its mother paralyzes and buries in a pit. The difficult problem is explaining how such complex and specific behavior developed in this species of wasp before the entire species became extinct.

Another major problem with the reductionist-materialist model is that it postulates that mind and consciousness—that are essential features of living organisms—*emerge* in them when they develop a certain level of complexity. However, this theory fails to explain both how and why these features evolved. The fact that they did evolve from matter implies that they are present there in *potential form*. If this were not true then it would imply that they magically arouse from matter—a substance lacking both mind and consciousness. Since all matter is composed of subatomic particles such as electrons and protons, then it follows that even an electron must possess the potentiality to express mind and consciousness. Of course, this is the argument of spirituality.

This question of why consciousness emerges in higher living organisms has also been posed by noted philosopher Thomas Nagel in his book, *Mind and Cosmos: Why the Materialist Neo-Darwinian Conception of Nature is Almost Certainly False*.[184] He questions why biologists are unable to explain why human consciousness and mental functions such as thought, intention, reasoning, and evaluation would confer an advantage to an organism that would be passed on to its offspring. These mental capabilities arose in humans, yet they utilize energy that might be used more productively for self-preservation. Natural selection is based on the success of the species in reproducing—complex mental abilities should play no role in this. Chance mutations can lead to more complex brains with increased mental capabilities or to less complex brains with less intelligence—the outcome cannot be predicted. Nor would there be any obvious advantage for survival of one outcome vs. the other. Biological evolution appears to proceed in the direction of increasing complexity, but there is no reason to conclude that such complexity confers a survival advantage—it could just as well be a disadvantage. Nagel concludes that neo-Darwinism's assumption that such capabilities arise from natural selection defies common sense.

The discussion about the creation cycle in Chapter 16 outlines how evolution is teleological. In other words how evolution is purposeful. This means that the evolution of species is not merely a result of chance mutations that led to selection of traits favorable to the survival of organisms, but that it also depends on the design or influence of cosmic mind. According to the model, the underlying goal or purpose of evolution is to produce sentient beings that have the capability of completing the final phase of the creation cycle—i.e. to attain union with cosmic consciousness. The spiritual model not only explains how life originated from inanimate matter but also solves the other problems that neo-Darwinism struggles to explain.

Summary

Materialist neo-Darwinism concludes that life, mind, and consciousness can evolve from matter without the help of an organizing principle. If this is indeed true then as neurologist Roger Sperry said, "There seems to be good reason to regard the evolutionary debut of consciousness as very possibly the most critical step in the whole of evolution."[185] According to the model complexity can arise spontaneously merely from a survival benefit. There is no question that living organisms have a survival instinct, and that no such trait is found in inanimate matter. But according to neo-Darwinism evolution depends on the ability of an organism to survive, but *the survivability of an organism depends on evolution.* This is circular reasoning and illustrates a fundamental flaw in the theory of evolution. By assigning the ability of matter to evolve into organisms that possess memory, intention, free will, self-awareness, etc., the neo-Darwinist is saying that inanimate matter can have the same properties found in living organisms. This is nothing more than restating the idea of primitive animism using scientific terminology.[186]

Spiritual ideology provides a simple and rational explanation of how life first arose on Earth and eventually evolved to produce human beings. It also explains how complex systems and organs developed, and why many traits that distinguish species appear to form in "quantum leaps" rather than slow, incremental steps. Using the analogy of quantum mechanics, we could postulate that many of the variations (mutations) needed to form a complex system such as the eye, lie in a state of potentia (uncollapsed superposition) until cosmic mind acts to collapse all the unexpressed gene variations at the same time thereby creating a large-scale change in the organism.

21

ANIMAL INSTINCTS AND HOMING ABILITIES

Instinctual Behavior of Animals

THE DEFINITION OF INSTINCTUAL behavior is behavior that is present in an organism without having been learned. According to the materialist model of biological evolution, there is only a single mode for inheritance of instincts—genes. While the science of genetics has made great strides in explaining how the physical attributes of living organisms are passed on between generations, genetics has yet to elucidate how complex instinctual behaviors in animals are passed from one generation to the other by their genes. DNA is the genetic material responsible for the transmission of the physical structures (cells) of all living organisms.[187] Theoretically, it can serve as the blueprint for constructing the entire organism starting from a single cell. This is known as genetic determinism—genes determine all biological form and function.[188]

DNA is nature's code for synthesizing proteins from constituent amino acids. Proteins are the basic building blocks of living tissues. They play a central role in biological processes. For example, enzymatic proteins catalyze chemical reactions and keep the machinery of life going. Others transport oxygen, move joints, run the immune system, and carry messages from cell to cell. Proteins are capable of performing many amazing functions and could even determine how the nervous system of an organism develops—even the microstructure of its neuro network and brain.

The model works for many of the basic animal instincts that can be traced to how the brain and nervous system develop, and to glands and certain structures and neural networks that affect specific behaviors or cause the organism to respond to stimuli in certain specific ways. For example, this is seen in our old friend the tiny roundworm *C. elegans* whose genome and connectome have been fully sequenced. Hence, genetics can explain simple autonomic responses and behaviors that promote self-preservation, but how complex instinctual behavior is passed from one generation to the next is a mystery.

Since DNA determines the sequence of amino acids that make up proteins any change in this sequence can change the structure and function of those proteins. But exactly how such changes could be translated into a different pattern of instinctual behavior is a confounding mystery. Take, for example, the instinctual behavior of a bowerbird. Bowerbirds have an extraordinarily complex courtship and mating dance. Males build an intricate bower decorated with sticks and brightly colored objects in an attempt to attract a mate. The bowers and courtship dance of all males of the species are similar, so this extraordinary behavior must be "hard wired" into the bird's brain since it is not learned behavior. In order for the behavior to be inherited, presumably it has to be programmed into the bird's DNA. This is the same as saying that there exists a genetic form of memory. The problem is that there is no evidence directly linking DNA with memory. DNA only codes for proteins. We are forced to assume that the bowerbird has a particular set of unique proteins that code for this complex behavior and other birds such as birds-of-paradise have different proteins that somehow code for their entirely different, but unique and elaborate courtship behavior.

If only biologist could find a repository for genetic memory in DNA, this would solve the problem, but no such storehouse is known. And if it were to exist, why wouldn't we inherit knowledge from our ancestors?

So the question is, what other genetic mechanism might explain unique and complex instinctual behaviors? Could physical changes in an organism during its lifetime be transmitted to its offspring? This is what the discredited hypothesis of Lamarckism postulated. If physical changes in organisms during their lifetime—such as greater development of an organ or a part through increased use—could be transmitted to their offspring then evolution could be a two-way street. Environmental conditions could affect heredity rather than the one-way flow of information in Darwinian evolution—i.e. only changes in genes could affect changes in offspring. However, today almost all biologists reject Lamarckism.

But a form of Lamarckism has been resurrected recently by biologists in the form of what is known as epigenetic inheritance. For this type of inheritance, changes in the environment do cause adaptive changes in the organism that can be passed on to subsequent generations. Although epigenetics violates Darwinian orthodoxy because it says changes in organisms' genes can occur because of environmental changes as opposed to mutations, biologists give it their okay because the inheritable features do not cause changes in the sequence of DNA—they are only due to the way genes are expressed. A classic example of this was how gypsy moths in London changed their color from brown to black in response to heavy air pollution during the Industrial Revolution. Epigenetics can explain how some adaptive changes in organisms might be passed on to subsequent generations, but most of these changes are neither permanent nor affect the entire species since the genome of the organism is not changed. Hence, it does not help to explain how complex instinctual behavior is passed from one generation to the next.

Although biologists have great difficulty in explaining complex inherited traits by genetics alone, spirituality has a simple solution to this problem. Animals have a connection to the nonlocal cosmic mind. Instinctual traits are archetypes for their behavioral patterns. Lacking the filtering mechanism that we possess that normally shuts us off from this collective mind, animals can tap into cosmic mind and are driven to behave according to the common archetypical patterns for their species. As new patterns of behavior emerge for the species that are adaptive, these behavior archetypes change. In this way, changes in a subgroup of a species can become incorporated into behavioral changes for the entire species. In other words, learned behavior of individual animals can theoretically be passed on to others, helping to advance the evolutionary development of the species.[189] This hypothesis solves the mystery of how animal instincts are passed from one generation to the next and helps to explain how species might evolve rapidly in response to environmental changes.

Homing Abilities in Animals

Many species of animals exhibit migratory and homing behavior, but exactly how animals navigate and find their way home is one of the great mysteries of science. Some animals migrate thousands of miles, yet they are able to return to a specific location on Earth with incredible precision. Scientists have posited several explanations for such abilities. These include the use of a magnetic compass based on the Earth's magnetic field; navigation using the sun; navigation using the stars at night; a keen sense of smell; and the use of mental landscape maps. A combination of such sensory cues might also be used. However, the magnetic field of Earth changes constantly over time and its position changes over the surface of the planet. It is also affected by sunspots, iron deposits, and magnetic anomalies and is so "noisy" that it is unlikely that it could be used to provide anything but a general direction to a homing animal. Sun navigation depends on both surface position and time. Animals would have to have an accurate internal clock to utilize this form of navigation. Mariners can obtain their latitude (distance from the equator) using the sun, a sextant, and an accurate watch. Alternatively, it can be measured using the pole star. But longitude (distance from the meridian at Greenwich, England) is not easily measured using the sun or stars. Therefore, navigation based on the sun or night stars can at best provide homing or migrating animals with a general direction—not a precise location.

The migrations of Pacific salmon are well known. After hatching, they live and mature in the gravel beds of freshwater streams for one to three years. Then the salmon make their way from their freshwater nurseries to the open waters of the Pacific Ocean, where they spend another several years feeding. Eventually the fish make their way back to the stream in which they were born in order to spawn and begin the cycle anew. One theory is that to find their way back across thousands of kilometers of ocean to the mouth of the river that leads to their home creek, the salmon imprint on the magnetic field that existed when they first entered the sea as juveniles. Next, a keen sense of smell might lead the salmon to their spawning grounds. The only problem is that we would not expect a stream to maintain a unique smell nor the Earth's magnetic field to remain the same during the several years between the time the salmon hatch and when they returned to their spawning beds.

Another example of a migratory mystery involves monarch butterflies.

They make a multigenerational, two-way, four-thousand-mile annual trip in which descendants of the third or fourth generation know exactly where the first generation started. Females deposit their eggs during the migration and die. The larvae become butterflies and somehow know where their parents were heading and make the second leg of the journey. But before completing this stage they die themselves and their offspring continue the migration. It takes four generations of butterflies to complete the cycle of migration begun by their ancestors.

Since, migratory and homing are instinctual behavior in animals, it is logical to postulate that they are guided by their connection to cosmic mind. This would also explain the mystery of how some lost pets find their way home.

Stories of Pets Finding Their Way Home

There are hundreds of reports of pets finding their way home after being lost. A couple of examples might illustrate why this is so difficult to explain from the materialist perspective that denies that animals are connected with a universal mind.

It was in November 2012 when a tortoiseshell cat named Holly jumped out of her family's RV in Daytona Beach, Florida, and ran off. After a fruitless search, the husband and wife returned home to West Palm Beach without their cat. However, in late December Holly showed up back in West Palm Beach, only a mile from their house. Because she had been micro-chipped, her identity was not in question, but how Holly found her way home across those 200 miles is a confounding mystery. Because of the worn condition of Holly's feet and claws and the loss of nearly half her body weight there is no doubt that she walked a considerable distance. With startling accuracy, Holly covered at least most of the 200 miles by prodding along day-after-day through habitat wholly unknown to her. Scientists cannot explain how she did it.[190]

During the 1920s, Bobbie, a two-year old female collie, was traveling with her family from Ohio to their new home in Oregon. During a rest stop in Indiana, Bobbie ran off and could not be found. After many hours of searching, the family gave up their search for Bobbie and continued on to Oregon. After three months, Bobbie turned up in the bedroom of their new home in Oregon when she awoke her owner by licking his face. Bobbie had never been to Oregon and there was no mistaking her because of several unique marks and scars. At the time, few believed the story that Bobbie had somehow made it half-way across the country to a home she never knew. The Oregon Humane Society launched an investigation and by interviewing people, they were able to reconstruct Bobbie's 2,800 mile trek home in the middle of winter by a different, more circuitous route than followed by her owners.[191] How could the dog find the new home of its owners unless it was connected mentally to the nonlocal cosmic mind?

Do Animals Have a Sixth Sense?

The herding of large animals, the flocking of birds, the schooling of fish, and the swarming of insects are well-known phenomena that undoubtedly have survival advantages for the species. What is interesting is how the individual animals are able to act in a concerted and collective motion that does not involve any central coordination. Watching a flock of starlings move in a rhythmic, wave-like dance is suggestive of a group mind at work. When the speed of the movements of a flock of birds is slowed down it turns out that they appear to react faster (0.015 sec) than their reaction time to a startle stimulus (0.038 sec).[192]

Rupert Sheldrake hypothesized that the uncanny ability of a flock of birds to react as though they were a single living organism is due to the telepathic power of the morphic field of the flock.[193] Such a nonlocal field connecting a group of living organisms would be through their connection with cosmic mind.

Throughout history, there have been reports of animals being able to detect hurricanes, earthquakes, tsunamis, and volcanic eruptions before they happen. Probably the earliest report was from 373 BCE in the Greek city of Helice, which described the flight of massive groups of rats, snakes, weasels and other animals, just days before an earthquake devastated the area.[194]

A more recent example was the report that before the giant tsunami hit the coasts of Sri Lanka and India on December 26, 2004 animals began acting strangely in the days leading up to the earthquake that triggered the devastating wave. Dogs refused to go outside, elephants trumpeted and ran for higher ground, flamingos abandoned their usual nesting areas, and zoo animals rushed into their shelters and could not be enticed to come back out. At the hard-hit Yala National Park in Sri Lanka, stunned wildlife officials reported that hundreds of elephants, leopards, tigers, wild boar, deer, water buffalo, monkeys, and smaller mammals and reptiles had escaped unscathed.[195]

There was at least one example in which authorities successfully forecast a major earthquake—based primarily on the observation of the strange behavior of animals. In 1975, Chinese officials ordered the evacuation of Haicheng, a city with one million people, just days before a 7.3-magnitude quake. Only a small portion of the population was hurt or killed because of the evacuation. It was estimated that the number of fatalities and injuries could have exceeded 150,000 if no warning had been given.[196]

Some animals seem to be able to sense that a person is about to die or have a medical condition or emergency. An example of this was reported in the July 2007 issue of the *New England Journal of Medicine* about a cat named Oscar. Oscar seemed able to "predict" the deaths of patients in a nursing home in Providence, RI. Just before patients died, Oscar would sit down by their beds and would become very upset if forced out of the room before the patient died. The article sited twenty-five successful predictions of impending death by the cat.

Finally, many dog owners have observed that their pet can anticipate their arrival back home. This was even studied scientifically by varying the time the owner returned and using alternate modes of transportation such as taxi.[197] Rupert Sheldrake also did an extensive study of such behavior that he published in his book, *Dogs That Know When Their Owners Are Coming Home*.[198] His studies of animal's psychic abilities indicated that animals also appear to know in advance when earthquakes, storms, tsunamis, and avalanches will occur.

Summary

In contrast to the reductionist model of reality, the spiritual worldview can offer a rational explanation for many of the mysteries involving animal instincts and homing behavior. This serves as just a small piece of the overwhelming evidence indicating that all minds are entangled with cosmic mind just as subatomic particles are entangled with one another. If the reader wants additional evidence for such a connection, I would recommend the excellent book entitled *One Mind* by Larry Dossey in which he details many such examples.[199] In addition, Dean Radin, chief scientist at California's Institute of Noetic Sciences, in his book *Entangled Minds* provides details of hundreds of experiment that demonstrate convincingly how minds can be connected.[200]

It is the argument of spirituality that our minds are connected through our mental connection with cosmic mind. Cosmic mind can be called the "Mind of God." It is postulated that not only our mind, but indeed, all minds mirror cosmic mind. And just like a hand-held mirror can reflect a small, but intense portion of sunlight, unit minds reflect the quality, but obviously not the full quantity of cosmic mind. In other words, all minds can be thought of as reflecting a portion of cosmic mind and it is through this that all minds are connected to the one mind.

We have also seen how consciousness is transformed into cosmic mind and the material world. One can equate God with cosmic consciousness, but unqualified consciousness is unmanifest and without the help of its co-existing creative principle (Sanskrit: *Prakriti*), it would remain so. Hence, consciousness might be inherently creative, but such creativity lies in potentia until the limitless cosmic consciousness allows itself to come under the influence of its creative principle. Why God would allow himself to become manifest is a mystery, but some sages have suggested that he might have been bored being unmanifest, and he decided in would be fun to witness a play—which they call his *liila*. Of course, we will never know the reason for why this creation exists, but here we are, and we are forced to deal with it.

According to spiritual ideology, the part of cosmic consciousness that becomes qualified is called *Saguna Brahma*—where *sa* means "with" and *guna* means "bondage." Similarly, the unmanifest cosmic consciousness is called *Nirguna Brahma* where *nirguna* means without quality. *Saguna Brahma* becomes the manifest universe, and the first phase of creation is the formation of cosmic mind with its progressively more qualified aspects of *I am*, *I do*, and *objective mind*. It is easy to see how our own minds are miniature versions of cosmic mind. We also function from the fundamental feeling of *I am*.

According to the spiritual model, cosmic mind is not static. It is a constantly evolving information source with limitless power to incorporate new information, meaning, and structure. Cosmic mind itself has no physical existence but a part of it is transformed into the material world. The model rejects the notion of classical deism that God created the world as a vast machine in time, and then sits back to watch it unfold. Instead, the Godhead is constantly creating our universe. The creation can be thought of as an "internal psychic" concoction of cosmic mind. Everything unfolds according to plan. We are only bit actors in the play but our role is to discover that *we are actually the one imagining the drama.*

The existence of the universal or cosmic mind explains the various anomalies and mysteries of physics, biology, psychology, and human longings for limitlessness. It is a wonder that so few scientists and intellectuals are aware of its existence and stick to their narrow view of reality. In the next part of this book, the emphasis will turn to the reasons for this and why the materialist worldview of reality is so harmful to human civilization.

PART FOUR

THE PROBLEM AND THE SOLUTION

22

WHY SCIENTISTS FIND IT HARD TO REJECT MATERIALISM

MATERIALISM IS A MORBID ideology that degrades human existence to little more than the accidental accumulation of elementary particles that move according to physical laws in an organized fashion only to disintegrate after a brief moment of geological time. By equating mind and matter, it essentially says human existence is meaningless with no higher purpose than to maximize pleasure. It is a worldview that envisions the universe to be a giant machine composed of separate and interacting parts. To quote Sir John Eccles, "It must be recognized that monist-materialism leads to a rejection or devaluation of all that matters in life."[201] In the face of the overwhelming evidence that contradicts this theory of reality, it is certainly pertinent to ask the question of why this undeniably false worldview is so popular.

The Problem with Science Today

Today the scientific worldview is primarily materialistic. This worldview dominates our culture and way of thinking about reality. People are naturally entranced by the many technological achievements of science. They look to science for the answers to all of today's problems. This attitude is reinforced by the numerous "talking heads" on TV that continually paint a picture that there is no need, nor any place, for spirit/consciousness in any discussion of the nature of reality and the numerous unexplained mysteries of the world. This means that consciousness is left out of almost all scientific discussions. Even in neuroscience, a discipline that should rightly be very concerned with consciousness, this important aspect of mind is for the most part ignored. One might expect consciousness to be a major area for research for scientists studying neurophysiology and the behavioral sciences—but nothing could be further from the truth. It is almost as though such studies are taboo and consciousness is dismissed as the hard problem. Anyone studying it must be either wasting their time or be a New Age kook. Part of the reason for this lack of interest in consciousness may lie in the fact that it does not fit the classical mechanistic, deterministic, local, reductionist model of science. By all measures, consciousness is nonlocal, nontemporal, nondeterministic, noncausal, and holistic. The "either/or" epistemology of classical science does not work when describing consciousness. It cannot be reduced to smaller parts. One needs to shift to the four-valued logic of "both/and" when describing it.[202] Few scientists are comfortable going there and those that are willing to brave such attempts risk irreparable damage to their professional reputations.

As mentioned previously, the prevailing method science uses to discover the secrets of nature is reductionism. Knowledge of reality is gained by breaking things down into their more basic parts. This has been a very successful method for understanding almost all aspects of the natural world. For example, studies in chemistry have shown how atoms bond to one another to form molecules, and how the structures of different molecules confer different and useful properties to them. There is no question that every aspect of our life has been improved by chemistry with its reductionist approach to understanding nature. The same could also be said for the advances in physics, medicine, molecular biology, electronics, materials science, etc. Given the success of the reductionist approach to understanding reality, it is not surprising that many scientists are reluctant to consider a different approach to understanding reality even in the face of evidence indicating reductionism does not work to describe the quantum realm, mind, and consciousness.

Reductionism assumes that subject and object are two entirely different things and that a subject's mind shouldn't have any effect on a physical object. The mental state of a scientist who performs an experiment shouldn't have an effect on the outcome of that experiment—the experiment should be reproducible no matter who performs it. Such an assumption is useful for doing science. For example, if a researcher is trying to discover a good drug for treating patients with the COVID-19 virus, it is best to assume that the mental state of the investigator or the doctor administering the drug does not affect its efficacy. Whether any particular drug is safe and effective for curing a disease should depend on its own merits, and not on the expectations of a medical professional or the prayers of the patient's friends and family. From a practical standpoint, such assumptions are necessary if science is to succeed in its mission of understanding the physical world. The problem is that this view has now become elevated to the level of absolute truth.

However, physics has revealed a flaw in the reductionist model. Can we really conclude that we have no effect on physical reality when *observation is required before any part of physical reality becomes manifest*? In fact, scientific discoveries indicate that we do participate in the way reality unfolds. A quote from physicist David Bohm illustrates what quantum physics tells us about reality:

> The inseparable quantum interconnectedness of the whole universe is the fundamental reality. And if we don't see this it's because we are blinding ourselves to it.[203]

But for the majority of physicists it is easy to become desensitized to the mind-numbing implications of quantum mechanics and just use it to do calculations for practical purposes. They merely accept the fact that nature is quirky, and go on with their lives and ignore the implications. This attitude is summarized by physicists Lederman and Hill in their book *Quantum Physics for Poets*:

> So bizarre are the consequences of quantum physics that, perhaps to preserve their sanity, the quantum physicist pioneers were driven to denial that they were actually describing a vast new reality, preferring to objectively insist that they had 'merely' invented a new method for making predictions about the results of possible experiments—and nothing more.[204]

In addition to reductionism, science today is largely mechanistic and works on the assumptions of determinism and locality. Like the reductionist approach, these are useful ways of doing science. Problem is that the evidence from quantum physics and studies of the paranormal indicate that determinism and locality are at best probable events—not absolute truth. In the face of overwhelming contrary evidence, most scientists take determinism and locality as scientific dogma. They have a deep-rooted sense that the world runs according to classical principles, and when it comes to doing science, it is a psychological challenge to shed the materialist mind-set. It is as though their psyche is not prepared to consider the fact that the quantum world is nonlocal and indeterminate—i.e. probabilistic.

Author Paul Levy in his excellent book, *The Quantum Revelation*, calls the impact of quantum mechanics on science's cherished classical perspective so shocking and discontinuous in nature that it has produced something in the physics community he calls "Quantum Physics-Induced Trauma."[205] Like any psychological trauma, physicists are slow to assimilate the mentally challenging revelations of quantum mechanics into their psyche. In the face of the unsettling discoveries surrounding the quantum, it is a natural reaction of scientists to cling to the mythos of what is comfortable—which in this case is the classical /materialist worldview. This trend may be slowly reversing, but for the most part scientists tend to stick to the old materialistic ways of thinking rather than embrace the alternate ideology of spirituality.

Rejection of Meaning

Scientific materialism rejects the idea that objective reality has any meaning. Meaning is a purely subjective construct according to this worldview. There needs to be a sharp line separating these two worlds—the objective world of physical reality and the subjective world of experience. When these two get confused or equated, materialists label it as insanity. The assumption is that we are the only ones capable of assigning meaning to a separately existing objective world.

Despite the fact that there is considerable evidence for the existence of a higher or collective mind for humankind that explains our common myths and archetypes, scientists tend to gravitate toward explanations for these phenomena that do not cast doubt on the dogma of separability. Science has become blinded by its almost axiomatic assumption that our inner and outer worlds of experience are separate and distinct. According to the materialist model, meaning could not possibly be universal. However, our experience is contrary to this. We create meaning with our minds and it takes shape in the physical structures and machines we design and build. In fact, spirituality takes the position that the whole universe is created from meaning—the all-encompassing cosmic mind. Since we are a construct of meaning, our life has meaning. To deny that life has meaning and purpose is to deny what is most important about us. It is not the intention of science to do this, but when it equates life with a machine, it inadvertently strips it of all meaning.

Matter-Based Reality and the Schism in Physics

Materialism is always looking for something that can function as the basis for physical reality. For example, Aristotle added the fifth element of heavenly aether to the four elements proposed earlier by Empedocles—earth, water, air, and fire. Alchemists attempted to purify, mature, and perfect certain materials, and aimed to transmute "base metals" such as lead into "noble metals" such as gold. They also attempted to create elixirs that might impart immortality and panaceas that might cure any disease. Behind all of these physical efforts, it is believed that their true aim was to achieve gnosis and a spiritual transformation of their mind.

More recently, science has looked to atoms and the subatomic particles that make them up as the fundamental stuff that constitutes physical reality. Then it was demonstrated that matter was an emergent property of an underlying, more fundamental realm, characterized by the wave function. Most scientists today are ignorant of what quantum mechanics implies about reality; and many of the physicists that have thought about the implications of quantum mechanics accept that there might be a more fundamental level of reality than matter, but work around the problem using their familiar reductionist model. Their attitude seems to be that if there is a deeper level of reality characterized by holism, it just means that scientific monism has come full circle. It's okay as long as we leave consciousness out of the discussion.

The only problem is that consciousness cannot be left out of any reasonable model of reality, and physics has as its primary mission the elucidation of reality. This causes a schism in the physics community.

On the one hand, physicists attempt to ignore consciousness because it is irreducible and unmeasurable while at the same time acknowledging that it is an important component of reality, without which physical reality could not become manifest. The Dalai Lama describes such a view of the world as schizophrenic.[206] Considering the fact that everything we know about physical reality comes from conscious experience, it is idiotic to try to factor it out of the theories used to describe reality.

The schism in physics is partly due to trying to fit a subtle, nonphysical reality into the materialist worldview. If the fundamental processes that govern how nature behaves lie outside spacetime, it means that this domain—call it the wave function, mind, or consciousness—is necessarily more fundamental than spacetime. This is a major clue that a basic assumption of materialism is flawed. Namely, that everything resides in the material realm of space and time, and should be measureable and observable within this realm—ultimately explicable by physical laws. Hence, the discoveries of quantum physics—particularly as they relate to entanglement and the unifying nature of the wave function—have put the materialist worldview in an impossible paradox. On the one hand, the ideology is locked into the concept of separate parts, three-dimensional space, and linear time, while at the same time acknowledging that behind the scene of spacetime lies a subtler, infinite, timeless realm of wholeness from which the constituent elements of physical reality emerge. It seems that few scientists today are even dimly aware of this contradiction and those that are, brush it off by saying that the ultimate nature of reality might be quite bizarre and mysterious, but just because we don't understand it doesn't mean we won't figure it out in the future.[207] The classical scientific paradigm has worked well in the past and there is no reason to think it won't solve this "little" anomaly about the nature of reality. Such scientists find no reason to think "outside the box."

Today most scientists are either ignorant of the problem with the materialist worldview or feel that questions about the ultimate nature of reality fall into the category of metaphysics and therefore religion—not physics. They feel that such matters are better left to theologians. Any evidence of the paranormal, such as psi phenomena, are readily dismissed as curious artifacts that arise because of people's wishful thinking, fraud, or are statistical aberrations resulting from poor scientific controls. They have bought into the dogma that nature does not work that way, and any evidence to the contrary must be false—a product of pseudoscience. It is easier to go with the mainstream and deny the veracity of paranormal phenomena than expose oneself to criticism from one's peers. And if there are concerns or doubts about whether the materialist worldview is correct, it is easy to compartmentalize such doubts in the same way that many scientists believe in God, regularly attend church, yet are advocates for the creed of scientific materialism.

People that gravitate toward science naturally have a questioning attitude and therefore a tendency to reject their religious upbringing. Many scientists that were brought up in a religious family setting now reject religion and strongly doubt the existence of God. As they gravitate toward a career in science, they become increasingly indoctrinated with scientific materialism. When given the opportunity, they tend to popularize the fantasy of the materialist worldview. In so doing, they think of themselves as defending science from superstitions and religious dogmas. The problem is that they fail both to recognize that materialism has its own dogmas and that spirituality is not religion. Although spirituality might be the foundation for all the world's great religions, it has none of the failings of organized religion.[208] The result is that in their attempt to purge science of what they consider religious "superstition," they close their mind to a more logical explanation that places consciousness as the fundamental ground substance of reality.

Perhaps the biggest obstacle for scientists to overcome if they are to reject the false doctrine of materialism is the unconscious conditioning they have received to remain part of the corporate/academic power structure that constantly reinforces the crudest manifestation of materialism—consumerism. Rampant consumerism permeates the economically privileged societies of today. It is characterized by an overriding concern for possessions, material wealth, and physical comforts. Profits, power, and control may be used to describe the global corporate power structure. Any higher human qualities are devalued. Scientists in both the corporate and academic spheres rely on funding from the corporate world and from the government—often for defense-related studies. Scientists are naturally discouraged from inquiring in directions that could threaten their livelihood, their salaries, research funding, and reputations.

The Skeptics

Skepticism has always been an important and healthy mindset for scientists. Science is based on evidence not beliefs. When the evidence suggests a paradigm shift is needed, it is natural to question the veracity of the evidence and demand a higher standard of it before overturning well-established modes of thinking or operating. And let's face it, embracing consciousness, instead of matter, as the fundamental "stuff" of reality would require a major paradigm shift for science. This would be a radical change from the customary way of viewing consciousness and mind. Even though it has repeatedly been shown that such a view is grounded in empirical science and spiritual ideology, it poses a major challenge to how science has been conducted for the last three-hundred years. Such a radical departure from the "normal" way reality is perceived naturally attracts much skepticism.

Skeptics are often heard to repeat the mantra that claims of an extraordinary nature require extraordinary evidence. Because of this, researchers of the paranormal have taken upon themselves to be very scrupulous in the controls and procedures they use in their experiments—much more so than other researchers in the life sciences.

Skeptics have gone out of their way to debunk anything that they considered pseudoscience—such things as acupuncture, astrology, chiropractic, ESP, faith healing, homeopathy, near-death experiences, reincarnation, etc.—in other words, anything that borders on the paranormal or is not currently a part of "main-stream" science.

Although it should be recognized that skepticism can be healthy, unfortunately this is only true if an open-minded approach to skepticism is used—not a dogmatic approach. The latter approach is known as scientism. It is a perversion of real science, an arrogant, materialistic tactic used to debunk anything a person believes is false. Here one pretends to have a questioning attitude and cites scientific jargon in an attempt to mislead people into disbelieving something that threatens that person's preconceived ideas. Lacking the ability to distinguish the difference between true science and scientism, much of the public is vulnerable to being led down a path that invalidates a spiritual explanation for reality when in fact it is nothing but the arbitrary and misleading opinion of persons making such claims. A true skeptic believes in the scientific method and must be genuinely curious enough to look at the data with an open mind before rendering an opinion. If they have questions about the veracity of the data then they acknowledge that they share some of the burden for establishing the proof or falsehood of what is claimed by other scientists. In this way, a true skeptic will detail the reasons for why they have a problem with the data and suggest ways to fix the problem.

Other Factors

Specialization in science makes it harder for people to consider an alternative to materialism. Science today is incredibly specialized. In order to be at the forefront of scientific research one must normally focus their energy on a very narrow slice of the pie. As a result, most scientists today have little inclination to study what might be considered philosophical questions about the nature of reality (metaphysics). It is natural to go with the majority opinion that matter is the ground substance of reality. To suggest otherwise leaves one open to scorn and possible harm to one's professional reputation.

Western religions have done little to dispel the myth of materialism. A majority of people in the West claim to believe in a Supreme Being that fathered the universe. But Judeo-Christian ideology rejects the concept of wholeness—teaching instead that physical reality and our separation from God and each other is part of the Divine plan. Other than the idea that we are all children of a Divine Father, people are not encouraged by theologians to consider the possibility that a deeper level of reality exists that connects everything and everybody. Hence it is no surprise that most of the public tend to follow the lead of scientists and the intellectual elite and fail to consider the possibility that consciousness is both the unifying principle and ground substance of creation.

Finally, rejection of the materialist worldview may require a more expanded consciousness. Unless a scientist acquires a wider, more universal or spiritual outlook on life, it may be difficult for them to accept an alternate ontology to materialism. An analogy would be a blind person who one day miraculously gained sight. That person would begin to perceive a far richer, more beautiful reality with its promise of a new meaning and purpose of life, opening up a new realm of almost infinite possibilities. Similarly, someone who is mentally fixated on the material realm may find it difficult to appreciate or tune into a nonphysical worldview. As a result, no amount of evidence is going to change their worldview—a change in how reality is perceived or experienced may be a prerequisite.

Summary

Humans have a truly enormous ability to ignore information that does not fit into their preconceived view of the world. Throughout history, we have witnessed this in the clash between religious dogmatism and science. Rather than look through Galileo's telescope to observe the moons of Jupiter, religious authorities shunned the opportunity and labeled his work 'that of the Devil.' And we have seen this in the dogmatic insistence that evolution is false because it is contrary to the biblical story of creation.

Throughout the history of science, we can also find examples of scientists who doggedly held on to preconceived attitudes about natural phenomena even when there was clear evidence contradicting their belief. For example, in the latter half of the eighteenth century the great majority of scientists were convinced that stones (meteorites) could not fall from the sky because from theoretical considerations they felt it was impossible for stones to form in the Earth's atmosphere. It was inconceivable to scientists at the time that any solid matter might exist in the heavens apart from comets, planets, and stars. Most reports of falling stones were simply rejected out of course, and the scientists that gave credence to such reports mostly adhered to the hypothesis that they did not originate from space, but instead were such things as volcanic ejecta or stones hit by lightning. This was despite the fact that there existed extensive collections of meteorites in several museums at the time.

Today few scientists seem ready to reject the materialist model of reality that has served science well for the last three hundred years. This is despite the revelations of quantum mechanics that indicate that there must be a deeper or subtler level of reality than matter. Such revelations are so shocking and disconnected to the classical mechanistic, deterministic, local, reductionist model of materialism that it can cause psychological trauma that forces most scientists to ignore the implications and seek out the more comfortable path of denial.

Evidence of the paranormal—such as psi phenomena—are equally destructive to this worldview, but again such evidence is easily ignored when it doesn't fit into one's preconceived ideas of how nature works, and accepting the truth of such phenomena might harm one's reputation and livelihood.

23

HOW MATERIALISM CONTRIBUTES TO SOCIETY'S PROBLEMS

Today's Problems

TODAY HUMAN SOCIETY IS characterized by social atomization, which brings isolation and feelings of powerlessness and unworthiness because collective approval is lacking. As human beings struggle with such feelings, the size of obstacles becomes exaggerated. They feel dehumanized, abandoned, hopeless, and unworthy. Under such circumstances, it is natural for many people to respond to the powerlessness and separation they feel with pointless acts of destruction—both of themselves and of others.

In America, the gap in wealth and income between the top 0.1 percent and the rest of the population has been growing for decades. In 1978, the group of the top 0.1 percent of the families (about 160,000) owned only 7 percent of the wealth, but by 2012, it had grown to 22 percent.[209] By 2020, the U.S. is approaching the same measure of inequality found in Russia. Wealth and income inequality undermines people's belief in progress and leads to a short-term view of the future instead of a more healthy long-term view. Fewer and fewer people believe that the future holds a better version of the present. Along with this, there is distrust for government and institutions because they appear designed to advance the interests of a privileged few

Technological change and globalization have displaced workers. Communities are destabilized by unemployment and the rapid movement of people. Family structure has disintegrated around the world. The global order has started to come undone as income inequality has increased and political ideologies have become more polarized, entrenched, and less willing to compromise.

An unregulated internet in the technologically advanced countries has made it possible for factually false ideologies and conspiracy theories to thrive. This has allowed radical, self-destructive, and conspiracy-based ideologies to gain a new popularity and have or threaten to spawn acts of terrorism and cause widespread destabilization of society. The psychological state of millions of people can be altered by bombarding them with misleading information and outright lies to influence how they vote in elections. For example, one needs only to look at the recent efforts of Putin's Russia to spread the false narrative that Russia did not invade Ukraine in 2014. Or their efforts to promote their candidate for the President of the U.S., Donald J. Trump, by hacking the servers of the DNC and Clinton's campaign manager and leaking selected emails as well as their unprecedented use of fake Facebook, Twitter, Instagram, YouTube, Tumblr, Reddit, 9GAG, and Google accounts to spread political fictions and paranoid phantasies that were simply untrue.[210] There can be little doubt that these efforts swung the 2016 election to Trump who has demonstrated his skill in using disinformation to further the ideology of nationalism and racism that is a hallmark of far-right ideologues and fascists.

Climate change, brought on by the indiscriminate burning of fossil fuels and the over reliance on animal agriculture as a source of food, threatens to displace millions of people, cause mass extinction of species, and rapidly alter the lands and waters that humankind depends upon for survival. In addition, there continues to be devastating insults to the environment by other human activities.

The healthcare system in America has been called, "fractured, unfair, inefficient, confusing, and anxiety-provoking as a result of its capture by a for-profit medical-industrial complex."[211] This has resulted in the skyrocketing cost of healthcare and the lack of universal healthcare in the country. As a result, the U.S. spends three times the amount of money per person as Great Britain, and has the highest cost of any country for prescription drugs. And even though the US spends more on healthcare than any other developed country, studies show that the additional money has not led to superior medical outcomes for Americans. The inability of many people in the U.S. to afford healthcare has brought on a health crisis, which itself leads to greater inequality.

This has contributed to the plague of suicides and opioid addiction that has so devastated large swaths of the U.S. causing untold deaths and suffering—especially to middle-aged white males.[212] Pandemics such as the recent one caused by the COVID-19 virus continue to pose a worldwide threat and are most devastating to the poor and to minority populations. Finally, 30 percent of scarce healthcare dollars in the U.S. go to "end of life" medical services, many of which are used to prolong the life of terminally ill patients that have negligible quality of life.

In many countries, a healthy plant-based diet has been supplanted by a diet rich in animal products. As a result, the total mass of domesticated farm animals is more than twice that of the mass of all human beings on the planet. Studies show that a diet rich in animal products is a leading cause of obesity, heart disease, and cancer.[213,214] Precious resources are lost converting food grains into meat for consumers in developed countries while at the same time creating a significant increase in greenhouse gases.

Communalism, provincialism, and nationalism are still rampant worldwide and are sources of conflicts, wars, and terrorism. Many people have a greater loyalty to an ethnic, religious group, or state than to society in general, and outnumber those that have a universal or spiritual outlook.

Religious fundamentalism continues to propagate an ideology of separateness in its efforts to reestablish archaic religious dogmas that deny the divinity of humankind. Even the mainstream religious institutions in the West affirm the reality of the material world and do little to criticize the accumulation of excessive resources by a few individuals while other members of society suffer from want.

The great social institutions of liberal education, democracy, and capitalism are beginning to crumble under the weight of the dehumanizing materialist worldview. The goal of a liberal education is to prepare individuals to deal with complexity, diversity, change, and provide them with a broad knowledge of the wider world (e.g. science, culture, and society). However, by emphasizing the material and ignoring the spiritual, it is failing to inspire a sense of universalism, social responsibility, and meaning in students.

Today democracy is directly threatened by increased inequality, as billionaires manipulate the media and finance candidates that will further their selfish interests. In the U.S., a corrupt campaign finance system allows a few very wealthy individuals with almost unlimited funds to essentially buy elections and exert disproportionate influence on the government and judiciary to the detriment of the general population. Also in the U.S., a defective presidential election system in which the residents of a few states get the privilege of choosing the president means that often the majority of voters are disenfranchised. In addition, the U.S. Senate has representational biases favoring less populated, rural states; and gerrymandering by state legislatures is done to establish an unfair political advantage for one political party vs. another. These contribute to inequality and the distrust people feel for democratic institutions. As democracy weakens, it gives way to oligarchy in which a rich and powerful few govern the many by invoking a political fiction of divisiveness, victimization, and gloom.[215]

Capitalism, the cherished economic system of the West, which is credited with bringing untold prosperity and social freedom, is beginning to crumble. Capitalism depends on a robust middle class to supply the capital and labor needed for economic growth, but growing income inequality is undermining this. Furthermore, an increasing world population and diminishing resources threaten to attenuate the rampant consumerism that is required to support this economic system. In addition, consumerism encourages the crudest form of materialism—a preoccupation with material wealth, physical comforts and considerations, along with a disinterest in cultural, intellectual, and spiritual values.

Today human society is plagued by pandemics, terrorism, mass murder, drug addiction, suicide, and numerous antisocial activities. People are either not concerned with or unaware that their actions might result in consequences in the future. A lack of knowledge and understanding of how everything is connected and the law of karma contributes to this scourge.

Why Materialism is a Root Cause of Many of these Problems

Materialism describes the separately existing objective physical world as reality. Implicit in this worldview is the understanding that this world is "*not I*." In other words, our subjective sense of *I* is separate and distinct from the outer world of material reality. Many scientists, intellectuals, politicians, economists, mainstream media, and capitalists have "bought" into the materialist worldview. As a result, the general public, which is burdened with numerous distractions, enamored by technological gadgets, and having limited time and education, has largely accepted the lead of these "elites " and accepted the core beliefs of materialism. But unfortunately most of the problems and conflicts that humanity faces today can be traced directly and indirectly to the acceptance of this ideology of separateness. When we buy into the idea that this objective world exists separate from ourselves, it necessarily distorts our own self-image and denies our connection with one another and the universe.

In addition, materialism today is often validated by scientism—the use of scientific jargon to promote one's own worldview. As Charles Tart aptly points out in his book, *The End of Materialism*, this leads to a sense of isolation for humankind:

> …most forms of scientism have a psychopathological effect on too many people by denying and invalidating the spiritual or transpersonal longings and experience that they have. This produces not just unnecessary individual suffering but also attitudes of isolation and cynicism that worsen the state of the world.[216]

The misplaced self-image that is created by the feeling of separateness can be likened to a "mind virus." Paul Levy calls this mind virus "Wetiko," after a Native American term for a self-replicating meme that is transmitted across generations.[217] Other societies give it other names, but to quote Levy it can be characterized as "a semantic disorder that functions by deviating the very process by which we attribute meaning to our experience."[218] He compares it to a collective psychosis that not only blocks information about reality but also substitutes a false narrative of what reality is. Being a mental disease, this false narrative infects the psyche of most people on Earth creating the highly damaging ideology of separateness. But because of its insidious nature it works primarily on an unconscious level and goes largely unnoticed. It is like a self-perpetuating myth that has been with us for a long time and will not go away on its own accord.

David Bohm recognized that a mental disease was infecting humankind and described it as follows:

> It's similar to a virus—somehow this is a disease of thought, of knowledge, of information, spreading all over the world. The more computers, radio, and television we have, the faster it spreads. So the kind of thought that is going on all around us begins to take

over in every one of us, without our even noticing it. It's spreading like a virus and each one of us is nourishing that virus.[219]

The mind virus of materialist ideology is dehumanizing since it deprives us of meaning. We become nothing more than sophisticated machines locked into an endless quest to maximize pleasure and minimize pain. This psychic disease is spread through all the various information and media outlets of the world. It is the primary reason that a large majority of people live on the surface of their being and are discouraged from diving deeper within themselves. To quote Levy, "To the extent that we're unconsciously identifying with a self-constructed model for who we are instead of recognizing and simply being who we are, we are living a lie."[220]

This mental disease is self-fulfilling in that the questions, perceptions, and experiments that scientists set up are designed to reveal the very fragmentation they believe exists in nature, and normally the results that come back merely reinforce this view. Fortunately, quantum physics for the first time has put a check on such thinking by showing that nature does not always respond in a fragmentary way when it comes to questions about the ultimate nature of reality.

Materialism proclaims that we have only one life to live. There is no afterlife, and there is no reason to be concerned with more than a "short term" vision of the future or that actions we perform might affect us after death. In other words, materialism denies that we might survive death in some form, and denies the existence of a karmic law that says that we reap what we sow. Since one's actions are divorced from their consequences there is no reason to act morally except for fear of being brought to justice by the court. People are thus ignorant that their antisocial behavior might echo back on them and that such behavior is actually not in their own best interest. Moreover, this puts pressure on relatives and the medical community to preserve life at any cost even when such efforts may be futile, costly, and do not produce a quality of life worth living.

Materialism functions by imposing social pressure on people to enjoy material objects. After all, if matter is the end-all and be-all of reality what other goal in life can there be besides the accumulation of material wealth? It promotes consumerism, which has quietly become the principle way in which most people measure themselves and others. The desire for material well-being becomes the goal of human existence. An artificial lifestyle is promoted that equates happiness with the accumulation of material objects. For those caught in the insidious web of consumerism, energies are directed toward the attainment of wealth, possessions, and social appearances, instead of the more gratifying pursuit of personal growth and being. The illusory value of consumerism is created largely by commercial interests and the entertainment industry. The emphasis on having rather than being is one of the factors that is responsible for the degradation of human society since it leads inexorably to selfish material desires rather than a concern for the well-being of society.

Materialism fosters individualism at the expense of self-sacrifice and service to others. From a biological and evolutionary perspective, there is nothing to be gained personally by service to those outside one's immediate family. Individualism leads to social atomization, serves to isolate people, and inhibits feelings of family and friendship for other members of society. As a result, people are unable to harmonize their diverse ideas and ideologies, and progress together, gradually transforming their self-interests into a unified rhythm, which is a characteristic of a healthy human society.

Materialist ideology considers the body to be a material object, and thus great emphasis is placed on allopathic medicine. The cost savings of alternative medical approaches such as homeopathy, acupuncture, biofeedback, meditation, prayer, chiropractic, naturopathy, Christian Science, and Ayurveda are not realized. The medical establishment has been schooled in reductionist ideology and for the most part believes that alternate approaches to healthcare have little, if any value. As a result, nontraditional and integrated medical treatments are ignored and a greater emphasis is placed on treating disease than on preventing it.

Materialism equates mind with brain. Hence, mind is created out of matter, and its existence as a nonphysical agency is denied. This implies that the mind has no business making moral or other judgments and that the search for meaning in life is a waste of time. By leaving consciousness out of the picture, materialism denies what is the most important thing about reality—it can only be experienced. Qualitative experiences such as love, beauty, awe, bliss, etc. are all we truly know. Materialism equates such qualia to our physiological responses to sensory experiences, which are mediated by hormones, or other chemical processes in the body.

Materialism not only devalues human life, but also animal life. Accordingly, animal life has no special value except as it might benefit humanity. This fosters the idea that the killing of animals for human consumption is morally justified even when doing so is not needed for our health or survival. Unfortunately, this attitude has a devastating impact on the environment.

According to a 2006 report published by the United Nations Food and Agriculture Organization (FAO), and a 2009 report by the Livestock and Climate Change environmental assessment experts at the World Bank, livestock and their byproducts account for the emission of at least 32,000 million tons of carbon dioxide (CO_2) per year. In addition, cows are responsible for the release of massive amounts of methane, a greenhouse gas that is approximately thirty times more potent than carbon dioxide, and which has reached record levels.[221,222] Together these represent 51 percent of the world's total greenhouse gas emissions and is greater than the transportation, electrical generation, and other industrial uses of fossil fuels combined.[223,224] In addition, it is estimated that the feeding of livestock uses over 30 percent of the Earth's entire arable land surface and uses vast amounts of scarce water resources. For example, it is estimated that it takes 2,500 gallons of water to produce one pound of beef, while the growing of feed crops for livestock consumes 56 percent of the water used in the U.S.[225]

Animal agriculture is also a leading cause of species extinction, ocean dead zones, water pollution, and habitat destruction. Cattle cause widespread damages to the environment in the form of land degradation and deforestation—especially in the Amazon River basin. Pollution caused by animal waste is a big problem. An example is the widespread pollution of rivers in eastern North Carolina caused by the overflowing of hundreds of hog waste lagoons that followed Hurricane Floyd in 1999, hurricane Matthew in 2016, and hurricane Florence in 2018.[226] In addition, the raising of livestock encourages the overuse of fertilizers and pesticides, and misuse of antibiotics—80 percent of which in the U.S. are used for livestock. This contributes to the creation of "superbugs" that are resistant to these drugs. Despite the great damage that raising animals for food causes to the environment, the U.S. government provides farmers and ranchers subsidies that encourage the overproduction of red meat and dairy.

Scientific materialism has been at war with religion for over five hundred years. The founding fathers of the U.S. were keenly aware of the threat that religious zealots posed for the young republic and were careful to allow people to practice their religion freely while insuring separation of church and state. As a result, secular and materialist ideologies are freely taught in the public schools while any teaching of spirituality is taboo since it could be construed as teaching religion. This is unlikely to change until science adopts a theory of reality that incorporates consciousness.

At the same time, people need to be better educated about spirituality and gain an understanding that it is not religion.

Summary

Scientific materialism has largely won the hearts and minds of most people living in the developed world. We might label it the reigning paradigm for reality. Many people who as children went to church regularly with their parents have since left the church and are now skeptical of the existence of God. As a result, they have gravitated toward the prevailing scientific ontology of separability, in which everything is derived from matter, and conscious existence ends when the body dies. Most of the respected scientists, doctors, intellectuals, business, and political leaders in the West subscribe to this doctrine. In its fight with religion to win over the hearts and minds of people, science has emerged the clear victor.

The problem is that materialism degrades the human mind by assigning a firm reality to the physical world and promoting an ideology of separateness. It denies people's connection to a higher reality. As a result, the majority of people live by the principle of separateness—living only on the lower planes of their existence. They identify themselves with their bodies and lower minds. In this state of ignorance, they see themselves as separate from the world and from other human beings. As a society, we pay a high price for subscribing to this doctrine. People erect social and psychic barriers between themselves and others—such as nationality, race, gender, religion, and economic status.

In the final analysis, materialism is no less dogmatic than religious fundamentalism. Both are faith based. For religion, it is the belief that scripture is the authoritative truth, while for science it is the belief that the natural world is derived from matter. However, the problem with dogma is that it is a preconceived idea that forbids human beings to go beyond its limits. It is a type of mental prison that one is expected to accept without question. It prevents the liberation of human intellect, and intellect is one of the greatest treasures of human beings, and it is a tragic situation when the human intellect cannot function freely.

What is most distressing is that materialism is a demonstrably false doctrine. The dogmatic attachment to this ontology is particularly damaging to society since it promotes the acquisition of physical objects, and weakens people's tendency to assimilate intellectual and spiritual ideas and values. Both the collective peace and individual peace are threatened by the erection of iron walls separating human beings from each other and from the natural world. Present-day materialism has brought humanity to this predicament, and its consequences are dreadful.

24

CHANGING SOCIETY FOR THE BETTER

Universalism

SPIRITUAL IDEOLOGY PROCLAIMS THAT every person, as well as every object in this universe, is a manifestation of cosmic consciousness. Human society is one, and each individual should be considered a member of a large family. In a joint family, every member is provided with the necessary food, shelter, clothing, education, and medical care needed for their survival—according to the economic resources of the family. If any member accumulates resources in such a way that it harms the other family members then they are chastised. On the other hand, if a family member is unable to contribute their fair share to the family because of mental or physical difficulties then the family will still accept that person with open arms and attempt to help them in any way they can. Human society must strive to do the same.

Many have tried to jeopardize the unity of the human race by creating factions. Such persons have a stake in creating divisions; they survive on the mental weaknesses of people and on their dissensions. Such people are afraid of the spread of an ideology of wholeness and exhibit their intolerance toward it in numerous ways, such as creating divisiveness, false propaganda, and lies. Knowledgeable people are not influenced by the dogma of separateness; they continually strive to perceive the unity of everything and everybody and think and act according to universalism and a new version of humanism, which are central aspects of the spiritual worldview.

A new version of humanism (neohumanism) was proposed by the Indian spiritual teacher Prabhat Ranjan Sarkar (1922-1990) to promote individual and collective progress. It is a holistic philosophical theory in which universalism plays a central role. Neohumanism follows directly from the top-down view of reality that describes everything as connected. Neohumanism discourages both geosentiment and sociosentiment, because both tend to be injurious to society. However, sociosentiment is particularly detrimental to society because it encourages one group of people to exploit a second group of people.

One of the principles of neohumanism is that all life has value—not just human life. Human life is especially precious because it is only with a human frame that one can attain union with cosmic consciousness. However, animal and plant life are of value and should only be destroyed when necessary. Neohumanism also counters the materialist idea that there is no higher meaning or purpose to human life. Consciousness is the universal substance of creation. It has no beginning or end. We human beings possess consciousness and our minds reflect cosmic mind. Thus we share in the immortality and limitlessness of cosmic consciousness.

How Adoption of the Spiritual Worldview Would Change Society for the Better

Materialism is taking humankind down the road to dystopia, but adoption of the spiritual worldview by a majority of people on Earth has the potential to take us down an alternate path to utopia. Spirituality describes everything as connected and numerous positive changes could occur if most members of society embraced this worldview. First, an understanding of spiritual ideology provides assurance that death of the body is not the end of our existence. People would no longer have a morbid fear of death. A new model for end-of-life care would emerge as people realize that death is a transition and not the end of their existence. Instead of spending a large portion of our limited healthcare resources on prolonging the life of terminally ill patients with questionable quality of life, loved ones and the medical establishment would be more open to letting people leave their body when they are ready.

Other aspects of the healthcare industry would change with the realization that the body has both physical and psychic strata, and that some illnesses are better treated using nontraditional and integrated approaches instead of the current "body-is-machine" mentality of allopathic medicine. The alternative approach of mind-body medical treatments is both cost effective and can often affect a real cure instead of just treating the symptoms. Moreover, as more people gravitate toward the spiritual path they naturally adopt a healthier lifestyle. They suffer fewer chronic illnesses. The scourge of obesity and drug addiction will be reduced, and universal healthcare will become more affordable as more emphasis is put on disease prevention as opposed to treatment.

Along with this new view of reality comes an understanding of the law of karma. We inevitably reap what we sow. People who understand how actions, either good or bad, inevitably return to affect them in the future have reason to act morally. That is, they are aware that there is no escaping the consequences of their actions. Moral behavior is in their self-interest. The result is a society with less crime and antisocial behavior.

Spirituality teaches people to dispense with the illusory sense of separateness caused by ego and endeavor toward the goal of becoming one with cosmic consciousness. People become more motivated to act in their true self-interest and discover the benefits of altruism, selfless service, and the sharing of scarce resources for the benefit of all.

Although human life is more valuable than animal life, if there is no need to take animal life in order to survive and be healthy, then it follows that it is morally indefensible to kill animals for food. A major benefit of a plant-based diet is that it reduces obesity and improves the health of individuals. Widespread adoption of a largely plant-based diet will help solve one of the greatest threats to humankind today—anthropogenic climate change. As mentioned in the previous chapter, over half the total worldwide greenhouse gas emissions can be traced directly or indirectly to animal agriculture. Cutting back on the consumption of meat will reduce greenhouse gas emission and help attenuate the effects of climate change. At the same time, we would enjoy enormous savings of scarce water resources. The reduction in the consumption of animal products as a food source will also reduce the untold suffering that is imposed on livestock, which is brought on by the mechanized assembly lines used to maximize the profits of the meat, dairy, and egg industries. An additional benefit of reducing the farming and consumption of animals will be the reduction of infectious diseases and pandemics most of which have crossed from animal to humans. Examples of such transmission include smallpox, measles, tuberculosis, 1918 Spanish flu, AIDS, Ebola, H1N5 bird flu, MERS, SARS, and COVID-19.

When people begin to realize that happiness comes from within rather than from without, they naturally feel less driven to acquire material wealth and possessions. Beginning in the 1980s, we have witnessed an unprecedented increase in human productivity due to the widespread use of automation, robotics, and AI. As intelligent machines take on more of the work of humans, productivity has increased and it has shifted jobs from the manufacturing to the service sector. This trend will continue and service sector jobs will undoubtedly be the next to go with the introduction of added and better AI. As a result, people will have increased leisure time, which will afford them the opportunity to spend more time on spiritual pursuits instead of the less satisfying pursuit of material goals. People will be able to find a new sense of self-worth without the need to spend almost half of their waking hours on a job.

As more emphasis is put on being as opposed to having, the sickness of accumulation of wealth, power, and fame will be gradually replaced by a healthy appreciation for the inner treasures obtained through spiritual practices such as meditation. The entertainment industry will also undergo a renaissance as people gravitate toward spiritually uplifting art, music, and other media as opposed to tales of violence and sex. As people develop awareness of the meaning and purpose of life, they begin to look up to spiritually elevated individuals. Individuals that acquired vast wealth or power by taking advantage of a flawed system or by exploiting people will no longer be idolized, and those with great physical prowess will no longer be held in higher esteem or be paid more than persons who display great intellectual ability.

Spiritual ideology is consistent with a socioeconomic system based on the welfare of the many as opposed to the few. Capitalism fails to provide this as it tends to concentrate wealth in the hands of a few families. An alternate socioeconomic system that ensures a more balanced distribution of wealth is that of P. R. Sarkar, called Progressive Utilization Theory or Prout. Prout would be a panacea for the integrated progress of human society. It aims to bring about equilibrium and equipoise in all aspects of socioeconomic life by totally restructuring economics.

Sarkar positioned it as an alternative to communism and capitalism. It recognizes the material world is common to all people and seeks the rational and equitable distribution of physical resources to maximize the physical, mental, and spiritual development of humankind. The Prout system of economic development seeks to guarantee everyone the five minimum requirements of life—food, clothing, shelter, education, and medical care. As an incentive, surplus physical resources are distributed to people who best serve and make the greatest contributions to society.

Prout is a type of progressive socialism that would restrict the accumulation of excessive wealth by a few individuals to the detriment of the many; advance cooperatives as the model for most businesses; and still incentivize people to be creative and productive.[227] A new socioeconomic system based on spiritual principles will go a long way toward insuring integrated progress in the economic sphere and the creation of a healthy human society.

Spiritual ideology encourages cooperation and service in place of selfish individualism that is so destructive to civilization. It teaches that service is good for the soul as well as for the person served. The service that most effectively dissolves the ego is that which is performed out of the goodness of one's heart, requiring nothing in return, not even recognition or thanks. Such service can be physical service, such as helping to build a shelter, or attending the sick; or economic service, such as relief work, feeding the poor, or helping the needy find a job. A higher form of service is intellectual service, such as teaching skills, general knowledge, morality, and spiritual philosophy. The highest form of service is spiritual service—performing spiritual practices. Such practices help the individual attain happiness, wisdom, empathy, and eventual unity. Others will want to emulate them due to their influence and example, and they will be drawn toward the spiritual path. Unlike physical forms of service, the effects of intellectual and spiritual service can be permanent in nature.

The material needs of people and other living organisms must be satisfied to maintain life, and life must be maintained to attain self-realization. All forms of true service weaken the grip of ego and bring one closer to the goal of unity. Hence, service advances our personal growth and produces a more harmonious human society.

To graduate from ego attainment to seeking union with cosmic consciousness is a desirable and healthy change that takes place as we age. However, for many people the transition never takes place. They remain engrossed in the crude pleasures of the body and physical world. They fail to learn of the untold benefits gained by searching within for peace and happiness and the answers to life's mysteries. In his book *Modern Man in Search of a Soul*, Carl Jung described the problems such people encounter as they age. They try to cling to the pleasures of youth as they enter the second phase of their life. They are accustomed to acquiring pleasurable experiences in the external world, but their sense organs inevitably become duller and their motor organs weaker. Such people seek greater and greater stimuli in order to compensate and may fall prey to what has been termed a "midlife crisis."

If left uncorrected, this attempt to stem the tide of old age can lead to an autumn of life marked by discontent, dissatisfaction, cynicism, and unhappiness. No matter how hard one tries, it is impossible to reverse the physical effects of aging. Jung argued that for the aging person it is vitally important to give serious attention to their personal psycho-spiritual development. The individual is freed from many of the mundane obligations of youth in their middle years, and they have the opportunity to reap the incredible treasures that an introspective approach to life can provide—for true happiness springs from *within*, not from without.

Knowledge of spiritual ideology will inspire more people to perform meditation and other spiritual practices. In a recent study, even a simple meditation practice was shown to cause significant positive changes in the brain. The study found that an average of twenty-seven minutes of the daily practice of mindfulness meditation produced a significant boost in gray matter density— specifically in the hippocampus—the area of the brain in which self-awareness, compassion, and introspection are associated. Furthermore, this boost of gray matter in the hippocampus was directly correlated to a decreased gray matter density in the amygdala—an area of the brain known to be instrumental in initiating fear, regulating anxiety, and stress responses. In contrast, the control group did not experience changes in either region of the brain, thus ruling out the possibility that the changes observed were due to the passage of time.[228]

Scientific studies have shown that meditation lowers blood pressure; lowers the level of blood lactate—reducing anxiety attacks; decreases tension-related pain and tension headaches; relieves ulcers, asthma, insomnia, muscle and joint problems; increases serotonin production—thus improving mood; and brings the brainwave pattern into an alpha state consistent with peacefulness, healing, and pain relief. Meditators report that they have increased energy; require less sleep—thus adding useful hours to the day; feel less stress; have improved metabolism; and can more easily lose weight. They feel more connected to other people and other life forms, and are happier than before they started meditating.

Other reported mental benefits of regular meditation include decreased anxiety; increased creativity and clarity, improved focus and peace of mind; development of intuition; increased self-confidence, self-awareness, and optimism; more harmonious relationships with friends, family and colleagues; and improved emotional steadiness and harmony.

The spiritual benefits of meditation constitute the true purpose for the practice. By focusing the mind on the here and now, it trains the mind to reduce the craving for the material things in life (e.g. money, power, fame, sex), which is responsible for destroying the contentment and satisfaction that comes from being rather than having. In addition, it leads to a personal transformation, as knowledge of one's true being obtained along with a deeper sense of purpose. This is accompanied by a happier more fulfilling life. Ultimately, the practice of meditation can result in the attainment of self-realization with its indescribable bliss, and the end to the cycle of birth and death.

An additional benefit of meditation practices is that they produce a love for the Cosmic Entity. All great saints and prophets had an overwhelming love for all manifestations of cosmic consciousness, and they radiated this love. Naturally, their charisma attracted many followers. Where there is love for the Cosmic Entity there is no personal ego, since ego is involved with the attachment for finite things. Devotional love is directed inward, as opposed to love for a spouse or child, which is directed outward. Arousing and attaining love for the Cosmic Entity is the goal but it is not easy. Yogis call such devotional love *bhakti*, and claim that this blessed state of mind is best obtained by practicing meditation, singing devotional songs, surrender, and selfless service. Devotional love for the Cosmic Entity creates complete trust in him and the assurance that all actions one performs are performed by him. Those persons who can develop strong love for the Cosmic Entity are well on their way to union with cosmic consciousness and are invaluable for inspiring others to enter the path of knowledge. The contributions that such people make to society are priceless.

Strong scientific evidence for the top-down ontology of spiritual ideology already exists. However, little funding is available for research investigating the awesome potential of the human mind and consciousness. If even a fraction of the resources put into building military weapons, discovering new subatomic particles, cancer research, or space exploration were put into the scientific research of mind and consciousness, then we might expect that the evidence would become overwhelming in favor of the spiritual worldview. A new paradigm of science would emerge that puts consciousness in its rightful place as the ground substance of creation.

The philosophical concept of wholeness would be incorporated into science in this new paradigm. The last vestiges of the dogma of scientific materialism would finally be demolished. Probably with the exception of a few remaining religious fundamentalists, people on Earth would have the intellectual freedom to explore spiritual ideology and learn of the great promise that it offers humankind. This change will eventually take place and will not only revolutionize science but also education, philosophy, ethics, and theology. Human society will enter a glorious new phase of development marked by harmony, peace, economic prosperity, political unity, and spiritual growth.

The Final Analysis

Today the tendency toward acquisition of physical wealth is extremely strong, yet humans long for limitlessness. When this longing for the infinite gets misaligned onto the material plane, people can develop an almost insatiable thirst for material possessions. They buy into the materialist worldview that their life ends with the death of the body and that their happiness depends on their economic status. If human longings for the infinite are allowed to run after objects of worldly enjoyment, conflict among human beings is bound to take place. As material wealth is limited, over-consumption for one leads to crippling scarcity for others. At the same time, people's interests and tendencies to assimilate spiritual or intellectual treasures become noticeably weakened. Separated by the iron walls of suspicion, people cannot trust each other.

All sense of humanity and all finer sensibilities are negated by materialism. The whole of human existence is degraded to its crudest foundations. Human beings gradually identify themselves with crude physicality or matter and as a result, their minds become cruder. The materialist worldview has created these problems for humanity and has contributed to the degradation of society.

The irony is that materialism is without question a false doctrine. When the light of scientific scrutiny is turned toward this false doctrine of reality, it is shown to be so transparent to its so-called scientific roots that it cannot even cast a shadow. There are so many inconsistencies and anomalies with this worldview that its proponents must take extraordinary pains to ignore the revelations of quantum mechanics and the evidence for paranormal phenomena. Its defenders seem to think they are defending science from the influence of New Age crackpots who want to introduce religion and superstition into science, but because of their unwillingness to consider the evidence for the veracity of the spiritual worldview, they are unwittingly abetting this false and degrading ideology. As a result, they are doing irreparable damage to society.

On the other hand, spiritual ideology holds the solution to the problems confronting the world today. Spiritual ideology is like the philosopher's stone that could transform everything into gold. Spiritual ideology holds the promise that when applied appropriately it can be used to find a just and rational solution to any problem. Unlike its supposedly "scientific" counterpart, spiritual ideology is totally logical and consistent with the scientific evidence indicating that there is a deeper or subtler level of reality—one of oneness—from which physical reality emerges.

Spiritual ideology confirms that what is true above is true below. In other words, the microcosm reflects the macrocosm. Ultimately, they are the same. However, we live in a relative reality that demands our attention. Although cosmic consciousness is unchanging and infinite, when it manifests as the creation it comes into the *realm of relativity* under the bondages of time, place, and person. We fall under these bondages and are relative entities that undergo constant change. Yet, we have a deep-seated longing for the infinite that can only be satisfied by allowing our being to unite with cosmic consciousness.

Everything in the material world is formed from consciousness and as human beings, we possess a developed mind and self-awareness. Our thirst for limitlessness can only be fulfilled through psychic and spiritual wealth. Consciousness, not matter, is the Source of all being and we must ultimately return to this Source in order to fulfill our ultimate destiny. In other words, every human being knowingly or unknowingly is on a quest for the Great.

Ultimately, the unity that we seek is attained by surrender. First the attachment to crude material objects is lost. Next our own individuality is surrendered. Oneness is experienced when we surrender our little *I* completely to the *Great I*. Such self-surrender is *not* akin to suicide. On the contrary, the individual soul will have its full expression. When we merge with God, we become God. Our existence does not become contracted but enlarged as we realize our true being is God.

The process is called enlightenment because it is awakening to a clear understanding of the wholeness that is Ultimate Reality. The word "awakening" is useful because it is analogous to awakening from a dream. While dreaming, one does not question the reality of the images that are witnessed, even though they may be quite bizarre. Upon waking up, we immediately dismiss the dream as unreal. Similarly, the illusion of separateness or nonunity appears real, but upon awakening to an enlightened state, one knows that the prior state of consciousness was illusory.

Readers may be convinced that this ideology, which holds the promise for a wondrous future for humankind, is logical and correct in its description of reality, but they should realize that an intellectual understanding of reality does not bring more than intellectual satisfaction. In order to experience the ecstasy of oneness and limitless happiness that exists in the infinite, unqualified cosmic consciousness, one must begin the difficult journey of self-discovery. Real change in human society takes place when the individuals that make up that society change.

APPENDIX:
A THEORY OF EVERYTHING

A SO-CALLED "THEORY OF Everything" (ToE) is the "holy grail" of physics. It would be a single all-encompassing theory that would explain everything about the universe. Finding a ToE is one of the major unsolved problems in physics. The two best theories upon which all modern physics rests are general relativity and quantum mechanics. But quantum mechanics only focuses on small-scale, non-gravitational forces while general relativity applies only to the macro scale. In general, these two theories are extremely accurate in making predictions in their domains of applicability, but are incompatible with one another in extremely small regions of time and space such as exist in black holes or during the beginning stages of the Big Bang.

The only way to resolve this incompatibility is to postulate a theoretical framework with a deeper underlying reality—the same way that the wave function explains quantum mechanics. Using this approach gravity might be brought into quantum mechanics. In pursuit of this goal, physicists have been actively researching what is called quantum gravity.

However, even if physics someday comes up with a single theory that unifies quantum mechanics and relativity, and successfully develops a new Standard Model that describes the four fundamental forces of nature—strong nuclear, weak nuclear, electromagnetic, and gravitational force—as well as all the observed elementary particles, it would still exclude consciousness. Would it be proper for physicists to say that consciousness is a phenomenon that is *not* part of the universe? I think the answer is obvious—any ToE must include consciousness.

In this appendix, I will attempt to outline many of the elements of a ToE based primarily on the spiritual ideology given by Shrii Shrii Anandamurti.[229] My chosen format is to answer questions that any ToE needs to address in order to be an all-encompassing theory that attempts to explain the unanswered questions about the nature of reality. These include questions about God, cosmology, quantum physics, relativity, paranormal phenomena, and the evolution and ultimate purpose of life in the universe.

Consciousness and the Geneses of the Universe

Can the existence of God be proven? Yes—if we equate God with consciousness. Then proof lies in the simple tautology that since we possess consciousness, it follows that God exists.

How did the universe begin? According to spiritual ideology, creation begins when the three binding forces of the cosmic creative principle (*Prakriti*) form an equilateral triangle and capture a portion of the unqualified cosmic consciousness (*Nirguna Brahma*) in the interior of the triangle.[230] The cosmic consciousness cannot withstand being confined in this manner and bursts forth from one of the vertices of the triangle beginning the process of creation. This explosion of consciousness from what is called the nucleus point of creation (*Purushottama*) is what initiates cosmogenesis—the Big Bang. The movement of consciousness, which begins in the cosmic nucleus, is extroversive or centrifugal in character and undergoes a change from subtle to crude. *Saincara* is the Sanskrit name given to this part of the creation cycle. This explains probably the greatest mystery of all time—what is the origin of the universe. Physics has struggled to explain how something could arise from nothing, but spiritual ideology explains this mystery very simply. Consciousness is transformed into cosmic mind and cosmic mind is transformed into spacetime, and then into the other fundamental factors. Hence, the Big Bang represents the emergence of spacetime from the cosmic nucleus, but it does not represent the true "beginning" of creation.

What came before the Big Bang? According to modern cosmogeneses theory, both space and time began with the Big Bang, but cosmologist have no idea of what occurred during the first 10^{-43} seconds following the initiation of the Bang. They call this period the "Planck epoch" and postulate that all the forces and quanta of the emergent universe were unified. According to spiritual ideology the unqualified cosmic consciousness burst forth from the nucleus of creation (*Purushottama*) but was under the influence of the creative principle and was first transformed into *I* am (*mahattattva*), then *I do* (*ahamtattva*), and then cosmic *objective mind* (*chitta*). In other words, cosmic mind is formed before spacetime (*akasha*), which is the first of the five fundamental factors that form during the creation cycle.

Is the universe infinite? No. Cosmic consciousness is infinite, but only a small part of unqualified cosmic consciousness comes under the scope of the creative principle, and it is known as *Saguna Brahma*. Therefore, *Saguna Brahma* is not limitless in the same sense as the unqualified cosmic consciousness (*Nirguna Brahma*). Astronomers are unsure whether the universe is flat and possibly infinite because the observable universe is limited by the speed of light. Because no signals can travel faster than light, any object farther away from us than light could travel during the age of the universe (estimated to be 13.8 billion years) cannot be detected, as the signals could not have reached us yet.

What drove the inflation of the early universe? The observable universe is approximately 93 billion light years in diameter. This is far larger than would be expected if it expanded at the speed of light for 13.8 billion years. In addition, the universe is extraordinarily smooth, which means that it is far more uniform than would be expected if it expanded chaotically. Inflation theory solves these two problems, but cosmologists have no idea what could explain the enormous energy required for inflation or why it stopped. Spiritual ideology solves this conundrum because it postulates that the source of inflation was the infinite energy contained in cosmic mind and it ceased when the creative principle gained the upper hand to stop the hyper-expansion of spacetime.

Is there a multiverse? No. Inflation probably does not lead to the creation of many universes. There is no need to postulate the untestable hypothesis of a multiverse to explain the incredible fine-tuning required for life to exist in our universe since all the physical forces, constants, masses, etc. were precisely set by the cosmic creative principle when the universe was created. However, there could be other universes if the same conditions that were responsible for the creation of *Saguna Brahma* occurred.

Is spacetime physical reality? Yes. Spacetime (ethereal factor or *akasha*) is one of the five fundamental factors that constitute physical reality. The others are aerial (gas), luminous, liquid, and solid. Although spacetime can be compared to nothingness or void, nonetheless it contains enormous energy, and after all, space occupies a volume and cannot be called nothing. Spacetime is also distorted by mass and can expand or compress according to the relative speed of an object. Yogis claim that spacetime emits a subtle vibration that can be heard in the mind as a sound. This is the *Om* sound, and they claim that it is the emergent sound of creation. We find some physicists waving their hands saying that the multiverse could arise from nothing. Their reasoning is that the universe could arise from nothing because it was a quantum mechanical fluctuation— just like the quantum particles that are observed to arise from the "nothingness" of spacetime and then disappear again.. However, since spacetime is *not nothing*, their reasoning is flawed. The real question should be where did spacetime come from? Spiritual ideology has a rational explanation for this.

How does spiritual ideology solve other problems facing the physics of the Big Bang? According to Big Bang cosmology all the mass-energy of the universe began from an incredibly hot and dense "cosmic egg" that was much smaller than an atom. Physicists call this a singularity. General relativity predicts that singularities can exist but the mathematics break down when they are used to try to describe such very small and heavy entities. Similarly, quantum mechanics cannot deal with singularities. This means that currently cosmologists have no idea of how the Big Bang was initiated and what took place in the very first fraction of a second following the "bang." Spiritual ideology has a solution for this problem. A portion of cosmic mind was first transformed into spacetime by the pressure of the static cosmic creative principle. It means that spacetime did not emerge from nothing. Initially spacetime began from a dimensionless point that expanded very rapidly. Concurrent to the formation of spacetime the aerial and luminous factors were created by the action of the static creative principle on spacetime. Initially the conditions in the early universe were so hot that matter could not form—there were just enormous amounts of extremely hot energy. After the hyper-expansion of inflation, the universe cooled enough for hydrogen and other light atoms to form.

What does the concept of entropy say about the origin of the universe? Physicists recognize that entropy has increased since the Big Bang and is the driving force behind the evolution of the universe. Increasing entropy means that the universe is constantly becoming less ordered. Turning back the clock of entropy means that the universe was in its most ordered state at the time of its birth. In fact, it could be argued that at the initial point of the universe—the Big Bang—it was in a state of complete orderliness (zero entropy). This would also be a way to describe cosmic consciousness. The order we witness in the universe today is a direct result of its initial order that can be traced to the Creator.

Will the universe end? No. Cosmologists predict that eventually all the hydrogen fuel for stars will become depleted and one-by-one the stars in the universe will run out of fuel and stop shining. Much later black holes will evaporate, and the universe will become a frozen wasteland having a constant temperature and no free energy to affect any change or movement. Such a thermal death of the universe assumes that no additional source of matter or energy enters the system other than that created in the Big Bang. Spiritual ideology counters this idea by pointing out that the universe is expanding. Along with the expansion of space-time, additional aerial factor (hydrogen) is constantly being created. This accounts for the fact that approximately one-third of the baryonic (ordinary) matter formed in the Big Bang is missing from galaxies. Recent astronomical observations show that this missing mass is found in strands of hot intergalactic hydrogen gas.[231] But astronomers don't know where it comes from. Creation is an ongoing process with additional space-time being created constantly along with small amounts of hydrogen gas. Therefore, a thermal or heat death of the universe will not occur. In addition, the theory that dark energy will continue to speed up cosmic expansion leading eventually to everything being so far apart that we won't be able to see other galaxies or even other stars in the Milky Way is wrong since it assumes that no new hydrogen gas is produced that can form new stars and galaxies.

What is dark energy and where does it come from? Dark energy represents the gravitational force of space itself. Originally, scientists believed in a static universe that neither expanded nor contracted. A repulsive factor for space was needed in order that the universe not collapse under the force of gravity but remained of fixed size. However, Hubble's discovery of cosmic expansion disproved the static universe model. Next scientists discovered that stars that are more distant are receding from us at an accelerating rate, not at all the rate that would be predicted if space was expanding regularly against the force of gravity because in this case, there should be a slight deceleration in the expansion. Cosmologists postulated that dark energy must be gravitationally repulsive and calculated that it accounts for roughly three-fourths of the total mass-energy of the universe. Subatomic particles are observed to continuously emerge and disappear into the "vacuum" of spacetime as it contains enormous energy. Spiritual ideology concludes that the observed expansion of the universe is due to the continuous nature of creation in which a portion of cosmic mind is constantly being converted into ethereal factor (spacetime), and a portion of spacetime is constantly being converted into hydrogen gas. Since real power resides in the subtle not the crude, spiritual ideology long ago predicted that the ethereal factor (spacetime) would have much more energy locked within it than all the other fundamental factors (gas, luminous, liquid, and solid) combined.

What is dark matter? In addition to the dark energy of spacetime, cosmologists also recognize the need for another mysterious substance that they call "dark matter." Calculations show that there is not nearly enough ordinary or observed matter in galaxies to hold them together. There must be a relatively massive amount of unseen matter in galaxies to account for their formation and stability. Estimates place the amount of dark matter at 20 percent of the mass-energy of the universe. This leaves roughly 5 percent of the matter of the universe as ordinary (baryonic) matter consisting of atoms and their constituent parts. Currently scientists have no clear idea what this unidentified matter is, nor does spiritual ideology offer any concise idea of its nature, but scientist's best guess is that it consists of weakly interacting massive particles (WIMPs).

Are there many-worlds? No. This interpretation of quantum mechanics is false and is contrived to explain some of the weird aspects of quantum mechanics (see also Chapter 4). It would be better to state this theory of reality as "many-possibilities." Physical reality emerges from cosmic mind, and cosmic mind is a subtle realm containing an untold number of possibilities—any one of which could become reality under the right conditions. For example, who hasn't imagined what would happen if they won the lottery or slipped off a high ledge of a cliff and fell? We imagine so many things that *could happen* to us, but almost never do. Standing outdoors with a thunderstorm nearby, we might worry about being hit by lightning, but fortunately, such events are rare. These are possibilities and reality unfolds according to our experience of it. According to many-worlds interpretation of reality, all such possibilities *do occur*—just in the other branches of the universe. Ridiculous when you think of it since we are only aware of the one universe we live in and experience.

Do we live in a participatory universe? No. This was the interpretation of quantum mechanics postulated by John Wheeler. He was correct in suggesting that quantum theory has opened the door to a new understanding of reality and that absent observers nothing could emerge from the realm of potentiality into physical reality. However, it is not logical to postulate that the consciousness of sentient beings is necessary for the emergence of the universe. It is a fact that the simple act of observing or measuring a quantum system will change the system. However, as pointed out by Bernard d'Espagnat, the objectivity of quantum mechanics is weak since the outcome of the observation of a quantum event is not dependent on *who* makes the observation.[232] It is more logical to postulate that this universe is the creation of cosmic consciousness and not that of an individual consciousness. If this were not so, then why do different unit consciousnesses obtain exactly the same results in experiments? Secondly, where in the hierarchy of conscious beings that might populate the universe would one draw a line and say this type of entity and not that one has the capacity to collapse a wave function? Spiritual ideology says that observership is indeed required to bring about physical reality, but the observer is the universe itself, not any particular part of it.

Is there any direct indication from physics for the existence of cosmic mind? Yes. In addition to equating cosmic mind with the domain of the wave function, there is a way to conceive of the universe as consisting of quantum black holes. Max Planck envisioned the three basic constants of nature to be the speed of light (c), the Planck constant (h), and the gravitational constant (G). By assigning a value of one to these three constants, Planck was able to come up with absolute units for time (5.3×10^{-44} seconds), length (1.6×10^{-33} cm), and mass (2.2×10^{-5} g). A quantum particle with these dimensions would be a tiny black hole that would flicker into and out of existence very rapidly. If we then assume that spacetime consists of these tiny, fluctuating black holes, then it would have an enormous density of 10^{94} g/cm^3, which is much greater than nuclear densities of 10^{14} g/cm^3. As a result, elementary particles represent an almost insignificant portion of the quantum field fluctuations of the spacetime vacuum. In other words, physical matter *cannot* be considered the basis for any physical understanding of reality—one needs to begin with this underlying realm. An analogy would be to compare physical matter to the foam that forms on the surface of the ocean as it is disturbed by wind and waves. The foam is insignificant compared to the ocean. Physicist David Bohn called this ocean the implicate order. We call it cosmic mind. Physicists ignore it because it cannot be measured.[233] Physical reality can be conceived as a manifestation of energy that is, as Bohm put it, folding and unfolding from this ocean of almost infinite energy. In other words, physics describes fundamental reality as a subtle domain containing enormous energy from which the ordinary elements of physical reality (including spacetime) emerge. This is also the description of reality described by the wave function and is consistent with spirituality's description of cosmic mind.

Is there any evidence from physics that matter is composed of consciousness? Yes. Most interpretations of quantum mechanics describe reality as unfolding from the realm of the wave function when it collapses due to entanglement with the more complex wave function of an instrument or observer. This is termed downward causation. Downward causation only works if at some basic level all matter is composed of consciousness. If this were not so, then an elementary particle like an electron would have no way of knowing that it must behave differently when we are observing it vs. when we are not observing it.

Is there any analogy in physics to mental time travel—e.g. precognition? Yes. although the laws of physics prohibits any material object from reaching or exceeding the speed of light, no such limitation would apply to movement that occurs at a more subtle level than spacetime. Consider for example an electron that undergoes a quantum leap between two energy levels in an atom. The electron is known to jump instantaneously without passing through spacetime. Such movement can be considered superluminal. Similarly, the quantum mechanical concept of tunneling allows matter to move between two points or events in spacetime instantaneously. Tunnels in spacetime are also known as wormholes, and they are allowed according to relativity theory. If a particle passes through such a wormhole, it is literally moving between two points or events in spacetime as though it was *out of this three-dimensional world.* Such movement could be between spacetime events or locations that correspond to either the past or the future. In other words, physics says we might gain information about past or future events as long as we "travel" not in spacetime, but in a subtler domain in which superluminal travel is possible—one which we could characterize as that of the wave function or cosmic mind.

What is the difference between the microcosm and the macrocosm? These are complementary ways of describing reality. Because of the unity or oneness of creation whatever exists above (macrocosm) also exists below (microcosm) and vice versa. For example, humans possess consciousness; therefore, the universe is conscious. The unit mind reflects cosmic mind, etc. The difference is generally in quantity not quality.

How Spiritual Ideology Explains Quantum Weirdness

What is the origin of physical reality? In Chapter 17, we learned how spiritual ideology explains the mysteries of quantum physics. One of the most important points of this discussion is that according to quantum mechanics, physical reality originates from the domain of the wave function, but this occurs only when there is an act of observation. In addition, this domain is essentially identical to that of cosmic mind. In other words, physical reality originates from cosmic mind, but consciousness is intimately involved in the process since the entire creation is the internal mental concoction of cosmic consciousness. An analogy would be for us to imagine a boy sitting under a tree reading a book. The image we construct in our mind is internal psychic to us and has no physical reality and as a result, others cannot share our mental construct. On the other hand, the internal psychic concoction of the Cosmic Entity appears to us as physical reality and we all share the same experience of it.

Which interpretation of quantum mechanics is most consistent with spirituality? Probably the Bohmian interpretation. Bohm's interpretation of quantum mechanics describes reality as having three layers. The crudest layer is physical reality including spacetime that he called the explicate order. The subtler level of reality that permeates our four-dimensional world of space and time he called the implicate order. This is identical to cosmic mind (or the domain of the wave function) and it unfolds into the manifest realm of reality—the explicate order. The third layer is the cosmic witness or consciousness (superimplicate order). He argued that ultimately, the entire universe (with all its particles, including those constituting human beings, their laboratories, observing instruments, etc.) has to be understood as a single undivided whole, in which analysis into separately and independently existent parts has no fundamental status. In other words, reality may be described as having three levels but this is only a useful way of characterizing reality, which is actually best described as whole, indivisible, with illusory parts that are intimately connected.

What does complementarity in the quantum realm imply about reality? Recall that the principle of complementarity was first formulated by Niels Bohr. His complementarity principle holds that objects have certain pairs of complementary or mutually exclusive properties that cannot both be observed or measured simultaneously. The best example is that of light, which has the dual aspects of behaving as a wave and as a particle. Neither description for light works under all experimental conditions. A complete description of light requires that we consider both of these mutually exclusive constructs, and our knowledge of the situation is limited because we are unable to simultaneously measure or describe both constructs precisely (uncertainty principle). Logically the existence of complementary aspects for describing a physical or mental construct indicates that the true reality of the construct must exist at a deeper level than that of the two complementary descriptions. Other examples of complementarity in the physical realm are space-time, position-momentum, and matter-energy. The concept of complementarity is also applicable to other fields besides quantum mechanics when there is no single or simple way to describe reality from our limited perspective. Examples of this type of complementarity are: particular-contingent, cause-effect, observer-observed, thought-action, physical-psychical, yin-yang, male-female, microcosm-macrocosm, part-whole, and relative reality-absolute reality. Spiritual ideology explains that all such complementary phenomena exist because the creation is a singularity (one). However, from our limited perspective, everything appears separate, but our experience of this separateness is illusory.

Why is there uncertainty in the quantum realm? Uncertainty exists in the physical and mental realms since everything in these realms has wave-like nature; and by definition, a wave is nonlocalized with indefinite position. Additionally, sense organs and instruments have limited precision, and the very act of physically measuring or observing an object affects it. Certainty only exists in cosmic consciousness since it has no wave-like character and has no movement or vibration whatsoever in time or space.[234]

Does determinism apply to physical reality? No. The ultimate limitations in precision described by the Heisenberg uncertainty principle and complementarity dictate that all properties and actions in the physical world manifest themselves non-deterministically (i.e. probabilistically) to some degree.

Why Relativity Theory is Consistent with Spiritual Ideology

What is block time? Einstein's theory of relativity leads inevitably to a new understanding of how space and time are inseparable. This new way of thinking of time is what is known as block time. The whole or entire four-dimensional continuum of spacetime includes all events that have or will take place. This means that our perception of the flow of time is illusory. From this new perspective of block time, things do not change in time.

What are the spiritual implications of block time? If all events that have or will occur are already present in the totality of spacetime, this means that beneath our ever-changing perception of reality lies a deeper, singular, and timeless reality. This realm has been identified as cosmic mind. During a mystical experience one may be able to perceive all four-dimensions of spacetime simultaneously—a capability of cosmic mind. Then that person could perceive events that happened in the past or will occur in the future. Such an experience might be difficult to describe. But a person might use phrases such as the stoppage of time, the experience of an eternal now, or a limitless, timeless realm to describe their experience. The experience would certainly lead one to the realization that our common perception of reality is but a shadow of a higher reality.

Spiritual Ideology's Explanation for the Paranormal

Do psi phenomena violate physical laws? No. Some scientists make the argument that psi phenomena are impossible because they violate the laws of physics. Albert Einstein made the same argument against the possibility of quantum entanglement—but he was dead wrong. Once it was proven to exist, physicists came up with a new understanding of the domain of the wave function to explain it. However, this explanation for entanglement can also be used to explain entangled minds and the existence of a subtle realm—cosmic mind—from which physical reality emerges.

How does spiritual ideology explain psi phenomena? Since physical reality is an emergent property of cosmic mind, it means this universal or one mind is connected with everything in physical reality. The transfer of information mentally can occur with no restrictions of time, place, or person—i.e. nonlocally. Since cosmic mind is universal, the mental impressions characteristic of telepathy, remote viewing, and precognition are merely information obtained nonlocally from this universal mind. The various psi phenomena are actually the same phenomenon—just different ways of characterizing information gained via the cosmic mind.

How does mind, which is nonphysical, affect physical reality? It is known that mental intention can collapse a wave function causing physical reality to manifest in a specific location or way.[235] This is the mechanism for how mind interacts with the brain. The brain is a quantum system. When we consciously intend to blink our eyes this mental intention causes the specific wave function for the nerves that control the eyelids to collapse in such a manner that the nerves are stimulated and the eyes blink. This is exactly the same phenomenon that occurs when we set up an instrument to measure the spin of an electron. Prior to observation, the electron is in a superimposed state of two possible spin states, but observation kicks it into one of the states. Similarly, the neurons in our brain exist in a superposition of possible states until our mental intention causes a specific outcome to take place. In other words, the nonphysical act of observation or intention affects the physical state of a material object. This is the mechanism for how mind affects matter, which also happens to be the very definition of psychokinesis.

If consciousness survives the death of the body, are stories of ghosts, fairies, angels, and demons real? No. Since the bodiless mind does not possess a physical or psychic body consisting of any of the five fundamental factors (ethereal, aerial, luminous, liquid, or solid), it does not possess any nerves or any other faculty that would allow it to affect a physical object. The unexpressed reactive momenta (*samskaras*) of a bodiless mind are all that differentiates it from other unit minds and from cosmic mind. The bodiless mind cannot find any outlet for expression until it finds a new physical body and is reincarnated. We may conclude that ghosts and other paranormal beings are not real.

Then what is behind such stories of paranormal beings? Are all such tales hoaxes or imaginary? To answer this question we must first consider the meaning of what is real and what is imagined, or a product of hallucination. The seat of the sense organs is in the brain and not in the gateway organs. For example, the eyes receive light and transmit nerve impulses to the brain where the actual sensation of sight occurs. This is true of the other four sense organs as well. Thus, everything we perceive actually occurs in our brain and is only indirectly dependent on the external world, which is what stimulates our gateway organs—eyes, skin, ears, nose, and tongue. Therefore, nerve impulses that affect our brain but have no link to the external world via our gateway organs can appear real. Such phenomena are often labeled hallucinations and certainly some experiences of ghosts fall into this category. Hallucinations are of two kinds: positive and negative. A positive hallucination occurs when thought waves affect the sense organs in the brain and one sees, hears, feels, tastes, or smells something that is not actually present in the external world. One's sense of reality is temporarily impaired, and the conscious mind may get absorbed into the subconscious mind. Unlike the dream state, whose reality we reject the moment we regain normal consciousness, the positive hallucination may seem real even after our conscious mind resumes normal activity. Positive hallucinations may also be elicited by hypnosis. Negative hallucinations on the other hand occur when the mind refuses to see something that is actually present. These can also be brought on by outer-suggestion (hypnosis) and by autosuggestion—most commonly when, because of fear, one's mind refuses to accept some aspect of an experience. Fear has a powerful effect on the mind. It can cause temporary concentration of mind. For example, if a person believes in the existence of ghosts and visits a house that is said to be haunted, fear may trigger a positive hallucination of a ghost. The concentration of mind caused by fear can leave an imprint in the cosmic mind at that location. That in turn may trigger the subconscious minds of other people and cause them to experience similar hallucinations. Therefore, most ghost stories are propagated by fear and have no physical reality.

    What about people's visions of sacred beings and divinities—are they real? Most visions of divine beings or personalities such as the Virgin Mary are produced by the psychic projections of individuals. Cosmic mind is the real creator of the universe but individual minds interpret and experience what cosmic mind creates. Some authors such as Michael Talbot have popularized the concept of a "holographic universe." He argues that individuals through their strong concentration of mind create holographic or ectoplasmic projections that others can witness and even photograph in the same way that ghosts may be created and witnessed by others. People may experience visions or feel they are communicating with "higher beings" or divine personalities but these beings have no actual reality. They can best be understood to represent the archetypes of religious deities that exist in cosmic mind and become translated by a person's psychological perspective.

What about luminous entities that are associated with ghosts and UFOs? There is another category of paranormal sightings that are explained by what in Sanskrit are called *devayonis*. These are luminous bodies, or advanced souls who were unable to attain liberation due to some strong desire or attachment that remained in their mind at the time of death. For example, suppose that a person meditated regularly for most of their life and they became spiritually elevated. In addition, suppose that during their life that person had a great love of the fine arts and in particular the intoxicating rhythms of fine music. At the time of their death, this person may lack the necessary devotion and fail to surrender their unit consciousness completely to cosmic consciousness due to their deep desire to continue to enjoy music. The mind of such a soul or unit consciousness may be too subtle to be reborn directly in a normal physical body consisting of the five fundamental factors. Instead, they take on a body having only ethereal, aerial, and luminous factors. Lacking a complete physical body with liquid and solid factors, they are unable to perform spiritual practices and continue in a normal manner on the path of self-realization. They will be able to enjoy the vibration of music sympathetically by frequenting concert halls and other places where music is played. However, until their desire (*samskara*) to enjoy music is satiated, which could take many years; they will remain as a luminous body and be unable to be reborn in a physical body. Sages have identified seven different types of *devayonis*, according to their different desires. Since such beings have a luminous body they can sometimes be observed by humans. Many sightings of angels, fairies, and friendly ghosts may in fact be *devayonis*. It is possible that some sightings of UFOs are luminous bodies, but most encounters with UFO's including abductions probably fall into the category of holographic projections produced by the mind of individuals.

Are there sentient beings on other planets in the universe? Yes. Sages confirm that other sentient beings exist throughout the universe. In addition, from a scientific perspective, it would be extremely likely considering the vast number of galaxies and stars in our universe that conditions similar to those that existed on Earth for the past 4.5 billion years would exist on countless other worlds in the universe. Therefore, intelligent life should be found on numerous other worlds.

Will or have we been visited by sentient beings from another world? Most unlikely. Because of the speed limit for light and the vast distances between star systems (measured normally in light years), travel to all but "nearby" stars is impractical. In addition, the conditions conducive to the formation of life are probably relatively rare, as is the requirement that those conditions last for a long enough time (several billion years) for intelligent life to evolve. These requirements probably insure that intelligent life within a galaxy is rare, and likely would be on planets or moons that are too distant to travel to in a meaningful period of time (less than a few hundred years). This also assumes that it would be possible to build a starship that could travel close to the speed of light. However, even though physical contact with alien civilizations is impractical there is no such limitation when it comes to mental communication. No doubt when human beings become more spiritually advanced in the future, they will acquire the ability to communicate mentally with distant civilizations—if they so desire.

Are occult powers real? Yes. It is well known that spiritual aspirants may develop certain psychic or occult powers as a result of their practices. These may include various abilities such as levitation, psychic travel, etc. These powers (*siddhis*) are not the goal of spiritual practice and may distract one from the goal, which is to become one with cosmic consciousness. The biggest problem with displaying or using such powers is that they may inflate the ego of the practitioner. People may be impressed by a person with such powers and begin to look up to and even worship that person as a great guru or prophet. Unless that person is fully realized like the Buddha, such admiration may lead to an inflated self-image, which is exactly the opposite of what is needed for success on the spiritual path. Another problem with using occult powers is that they inevitably make the mind cruder and are eventually lost if used continuously. The power to heal the sick may seem beneficial; however, it can involve taking another person's reactive momenta onto oneself, and can potentially rob the sick person of a needed life lesson as well as adding karmic burden to the healer. Because persons who use spiritual powers can potentially do great harm to society, spiritual teachers (gurus) in the past were very careful to take only those students that were morally elevated and had a genuine yearning and love for God.

The Question of why there is Good and Evil

Why if God is Good is there evil in the creation? Although Western religions have struggled with questions about good and evil and suffering, spiritual ideology has a simple explanation for these. Good action is an action performed with good intention and knowledge—an action that brings one closer to the goal of becoming one with God. Evil is just the opposite and is performed out of ignorance. In the simplest sense, good is the movement from crude to subtle, or the process of identifying with the Cosmic Self as opposed to the ego. The Sanskrit term for this is *vidya*. Evil action may be termed *avidya*, or the movement toward ignorance or crudity. Such actions create negative reactive momenta, which lead inexorably to pain and suffering. If sentient beings are to have free will then necessarily they may make choices that on the surface appear as evil actions. However, individuals inevitably learn that such actions bring pain and suffering and eventually all humans will attain union. The Cosmic Playwright has set up all the physical and mental laws (rules) governing the creation, and we have no choice but to follow them.[236]

The Ultimate Merger with Cosmic Consciousness

What is a seeker? Human beings have a deep connection to cosmic consciousness. They have an inner voice that speaks to them in subtle whispers, and this voice is only quieted by the constant activity of the conscious mind or the stupor of unconsciousness. When an individual finally comes to the realization that there is more to life than the incessant quest for pleasurable experiences and achievements, then they inevitably begin the path to spiritual union. Such a person can be called a "seeker." A seeker has acquired wisdom and come to the realization that knowledge of self is knowledge of God. Such a person rejects the idea that extroversive activities bring lasting happiness. They have come to realize that true happiness comes from within. A seeker has begun the search for the meaning of life and the attainment of self-realization. The major concerns of the ego are put aside as the seeker begins to give themselves to selfless service and develops feelings of universal love for humanity. The seeker has an understanding that there lies a far deeper and richer reality beyond ego attainment and the experiences of the sense and motor organs. In a sense, a seeker is someone who has crossed the threshold from the stagnant waters of a pond into the flow of a stream that will eventually lead to the ocean. Seekers are characterized not only by an interest in attaining self-knowledge but also by their practice of some form of introspection such as meditation. In their search for knowledge, such individuals become tolerant, nonjudgmental, loving, selfless, service minded, more in tune with their body and nature, healthier, and happier than they ever thought possible.

What is the next stage after seeker? Slowly the seeker is transformed into a person that can be called a "seer." Such a personality feels love for every living creature and every particle composing the universe. They are completely open, no longer play psychological games, and are incapable of feeling any emotion except unconditional love for God and his creation. They live totally in the now and have access to the unlimited knowledge that lies in cosmic mind. The seer is not attached to the fruit of their actions; they feel that their actions are those of the Cosmic Entity. They create no new karma and know with certainty that when they leave their physical body they can become one with cosmic consciousness. A seer is completely free, living in a state of grace and indescribable bliss. It is natural for people to be drawn to such individuals and look up to them for advice and spiritual guidance. Knowingly or unknowingly all human beings crave the limitlessness that accompanies this, the ultimate phase of life.

How can one act and not produce additional karmic burden? Surrendering the ego is a technique designed to reduce our karmic burden and reduce our sense of separation and to increase our love of God. This technique is taught by all the world's great religions. If an action is performed with the ideation that God is performing the action, then the mind suffers no reaction to that action. The thought can be, "I am the machine, and God is the machine operator." The Sanskrit term for this type of ideation is *madhuvidya*, which means ascribing godhood to every living organism and object. It is a practice that allows one to carry on a normal worldly life and not create additional karma. Selfless service is also a great tool for diminishing the ego's control over our life and allowing us to see through the illusion of separateness. By serving others as though one were serving God, our exterior and interior is filled with cosmic bliss and all afflictions are extinguished.

Why should we try to transcend ego consciousness? The experience of cosmic consciousness is the ultimate experience, but it is an egoless state. This state of union with cosmic consciousness is called many names including self-realization, enlightenment, liberation, salvation, nirvana, samadhi, satori. In such union, we bask in the ecstasy of pure being. In reality, humans are no more separate from the Source than rays of sunlight are separate from the sun, but ego gets in the way by creating the illusion of separateness. Ego obscures Spirit, covering it with layer upon layer of "I am so and so," and "I do such and such." Ego is the source of our arrogance and suffering, and to believe that it should be empowered is the epitome of ignorance. Most people are unaware that there are practices that can speed up their spiritual growth and end the nightmare of constant rebirth. They continue living life on the surface of their being. Inevitably, they learn from the struggles of life experiences that it is necessary to subjugate the ego to attain true happiness. Often it is hardships and feelings of unhappiness, dissatisfaction, pain, and suffering that cause people to change the direction of their lives. Hence, suffering can be a blessing in disguise, and great suffering can beget great growth. This process entails the burning of negative reactive momenta, which brings mental anguish, pain, and suffering, but in the process the ego becomes weaker, yielding more of the sentient *I feeling*. For some people these experiences are what are needed to turn them away from the false promise of ego attainment—"the material path"—to the path of unity.

What is the final step in attaining unity? Intellectual knowledge is known as *jinana* is Sanskrit. Unfortunately, intellectual knowledge of spirituality is no substitute for spiritual experiences. Knowing in our mind that we are God does not create the feeling that we are God. Intellectual knowledge is useful for educating others about spirituality, but it has limited benefit for a spiritual aspirant. Its greatest value lies in pointing a person toward the path and getting them started performing spiritual practices. At worst, it can create egoism because the person believes they possess great wisdom, and should be recognized for this. This attitude can be an obstacle for advancing on the path. All the great sages have taught that God is best known through love. Spiritual practices are designed to unlock our unconditional love for God. Sanskrit has a specific word for devotion or love of God—*bhakti*; and the highest form of yoga is known as *bhakti* yoga. In the beginning, the spiritual aspirant may feel that his or her Supreme Father is always with them and both feel and see their Father everywhere—within and without. Next, the spiritual aspirant may develop unconditional love for God and feel blissful and will want to share their bliss with others through their love and selfless service. In the final stage of *bhakti*, the spiritual aspirant feels they and their Lord are the same. The aspirant enjoys the indescribable supreme bliss (*ananda*) from entering into the limitless ocean of cosmic consciousness. In this state, there is not even a touch of duality and the devotee forgets themselves and all their little predicaments—their unity with God is complete. They become fully ensconced in the ecstasy of oneness with God. This state could last for a moment or for the rest of their life.

BIBLIOGRAPHY

Agar, W. E., and F. H. Drummond. "Fourth (final) Report on a Test of McDougall's Lamarckian Experiment on the Training of Rats." *Journal of Experimental Biology,* 31, 1954.

Alexander, Eben. *Proof of Heaven: A Neurosurgeon's Journey into the Afterlife.* New York, NY: Simon & Schuster, 2012.

Anandamurthi, Shrii Shrii. "Matter and Spirit," Published in *Ananda Marga Ideology and Way of Life,* Part 5, 1971.

Anderson, Kurt (2020). *Evil Geniuses: The Unmasking of America: A Recent History.* New York, NY: Random House, 2020.

Behe, Michael. J. *Darwin's Black Box: The Biochemical Challenge to Evolution.* New York, NY: Free Press. 1996.

Belluck, Pam. "A Cat's 200-Mile Trek Home Leaves Scientists Guessing," *New York Times,* January 19, 2013.

Bohm, David. "On the Intuitive Understanding of Nonlocality as Implied by Quantum Theory," *Foundations of Physics,* **5,** 1975.

Bohm, David. *Wholeness and the Implicate Order.* New York, NY: Routledge & Kegan Paul, 1983.

Bohm, David. *On Dialogue.* New York, NY: Routledge, 2007.

Bourzac, Katherine . "Methane Emissions Reach Record Levels," *Chemical and Engineering News,* July 20, 2020.

Bowman, Carol. *Children's Past Lives: How Past Life Memories Affect Your Child.* New York, NY: Bantam, 1997.

Bowman, Carol. *Return from Heaven.* New York, NY: HarperCollins, 2001.

Bruce, Robert, and Brian Mercer. *Mastering Astral Projection: 90-day Guide to Mastering Out-of-body* Experience. St. Paul, MN: Llewellyn, 2004.

Bucke, Richard M. *Cosmic Consciousness: A Revolution in the Study of the Human Mind* . New York, NY: Dutton, 1969.

Campbell, Joseph. *The Hero with a Thousand Faces,* 3rd edition. Novato, CA: New World Library, 2008.

Carhart-Harris, Robin, et al. "Neural Correlates of the Psychedelic State as Determined by fMRI studies with Psilocybin," *Proceeding of the*

National Academy of Sciences, February 7, 2012.

Carroll, Sean. *The Big Picture: On the Origins of Life, Meaning, and the Universe Itself*. New York, NY: Dutton, 2017.

Chalmers, David. *The Conscious Mind: In Search of a Fundamental Theory*. New York, NY: Oxford University Press, 1996.

Clark, Nancy. *Divine Moments: Ordinary People Having Spiritually Transformative Experiences*. Fairfield, IA: 1st World Publications, 2012.

d'Espagnat, Bernard. *In Search of Reality*. New York, NY: Springer-Verlag, 1983.

Davies, Paul. *In Who We Live and Move and Have Our Being: Panentheistic Reflections on God's Presence in a Scientific World*. Editors: Philip Clayton and Arthur Peacocke. Grand Rapids, MI: Eerdmans Publishing, 2004.

Dawkins, Richard. *The God Delusion*. New York, NY: Mariner, 2008.

Dossey, Larry. *Recovering the Soul: A Scientific and Spiritual Search*. New York, NY: Bantam, 1989.

Dossey, Larry. *Healing Words: The Power of Prayer and the Practice of Medicine*. New York, NY: HarperCollins, 1993.

Dossey, Larry. *One Mind: How our Individual Mind is Part of a Greater Consciousness and Why it Matters*. New York, NY: Hay House, 2013.

Dunne B. J, and R. G. Jahn, R. G. "Experiments in Remote Human/Machine Interaction, *Journal of Statistics Edu.* **6**, 1992.

Eccles, John C. *Facing Reality: Philosophical Adventures by a Brain Scientist*. New York, NY: Springer-Verlag, 1970.

Efron, Robert. "Biology without Consciousness—and its Consequences," In: *Logic, Laws and Life: Some Philosophical Complications*, ed. R. G. Colodny. Pittsburgh, PA: Pittsburgh University Press, 1977.

Fein, Y., et al. "Quantum Superposition of Molecules beyond 25 kDa," *Nature Physics* volume **15**, 2019.

Fenwick, Peter, and Elizabeth Fenwick. *Past Lives: An Investigation into Reincarnation Memories*. New York, NY: Berkley, 1999.

Feynman, Richard. *Feynman Lectures on Physics*, vol. 3. Reading, MA: Pearson-Addison-Wesley, 1963.

Ford, K. "Update on John Archibald Wheeler" (PDF*). Princeton Physics News*, **2** (1), 2006.

Gefter, Amanda. *Trespassing on Einstein's Law*. New, York, NY: Bantam, 2017.

Gerrig, Richard, and Phillip Zimbardo. *Psychology and Life*, 20th ed. Essex England: Pearson Education, 2014.

Goodland, R., and J. Anhang, J. "Livestock and Climate Change: What if the Key Actors in Climate Change were Pigs, Chickens, and Cows?" *Worldwatch*, November/December, 2009.

Goswami, Amit. *Creative Evolution: A Physicist's Resolution Between Darwinism and Intelligent Design*. Wheaton, IL: Theosophical Publishing House, 2008.

Greene, Brian. *Until the End of Time: Mind, Matter, and Our Search for Meaning in an Evolving Universe*. New York, NY: Alfred Knopf, 2020.

Gribbin, John, et al. *Q is for Quantum: An Encyclopedia of Particle Physics*. New York, NY: Simon and Schuster, 2000.

Guerrer, Gabriel. "Consciousness-related Interactions in a Double-slit Optical Interferometer," *Open Science Framework*, March 9, 2018.

Harris, Sam. *Waking Up: A Guide to Spirituality without Religion*. New York, NY: Simon & Shuster, 2014.

Hawking, Stephen, and Roger Penrose. *The Nature of Space and Time*. Princeton, NJ: Princeton University Press, 1996.

Hawking, Stephen ed. *A Stubbornly Persistent Illusion: The Essential Scientific Works of Albert Einstein*. Philadelphia, PA: Running Press, 2009.

Heisenberg, Werner. *Physics and Philosophy: The Revolution in Modern Science*. Amherst, NY: Prometheus, 1958.

Herbert, Nick. *Quantum Reality: Beyond the New Physics*. Garden City, NY: Anchor Press/Doubleday, 1985.

Hitchens, Christopher. *God is Not Great: How Religion Poisons Everything*. New York, NY: Hachette, 2007.

Holt, Lester NBC Nightly News on March 20, 2015.

Hölzel, Britta K, et al. "Mindfulness Practice Leads to Increases in Regional Brain Gray Matter Density," *Psychiatry Research: Neuroimaging*, **191**, no. 1, 2011.

Horowitz, Julianna M., et al. "Trends in Income and Wealth Inequality," 2020.

Howe, Quincy Jr. *Reincarnation for the Christian*. Wheaton, IL: Theosophical Publications, 1974.

Huxley, Aldous. *Doors of Perception*. New York, NY: Perennial Library, 1990. (Originally published 1954).

Jabr, Ferris. "How brainless slime molds redefine intelligence." *Nature*, 13 Nov. 2012.

Jahn, R. G., et al. "Correlations of Random Binary Sequences with Pre-Stated Operator Intention: A Review of a 12-Year Program Journal of Scientific Exploration," Vol. 11, No. 3, 1977.

Jahn, R. G. "The Persistent Paradox of Psychic Phenomena: An Engineering Perspective," *The Proceedings of the Institute of Electronic and Electrical Engineers*, **70**, no.2, 1982.

Jahn, R. G., B. J. Dunne, and R. D. Nelson. "Engineering Anomalies Research," *Journal of Scientific Exploration*, **1**, 1987.

James, William. *The Varieties of Religious Experience.* New York, NY: Mentor, 1958. (Original work published 1902).

Jung, Carl G. *Man and His Symbols.* New York, NY: Bantam. 1964.

Jung, Carl G. *The Symbolic Life.* Princeton, NY: Princeton University Press, 1976.

Kahleova, Hana, et al. "Effect of a Low-Fat Vegan Diet on Body Weight, Insulin Sensitivity, Postprandial Metabolism, and Intramyocellular and Hepatocellular Lipid Levels in Overweight Adults." JAMA Network Open, 3 (11), 2020.

Kelly, Edward. F., et al. *Irreducible Mind: Toward a Psychology for the 21st Century.* Lanham, MD: Rowman & Littlefield, 2007.

Koch, Christof. "Proust among the Machines." *Scientific American*, December, 2019.

Kothari, L.K., et al.. "The Yogic Claim of Voluntary Control over the Heart Beat: an Unusual Demonstration." *J. American Heart Assoc.*, **86**, no. 2, 1973.

Langley, Noel. *Edgar Cayce on Reincarnation.* New York, NY: Warner Books, 1967.

Lederman, Leon M. and Christopher T. Hill, Christopher. *Quantum Physics for Poets.* Amherst, NY, Prometheus Books, 2011.

Levy, Paul. *Dispelling Wetiko: Breaking the Curse of Evil.* Berkeley, CA, North Atlantic Books, 2013.

Levy, Paul. *The Quantum Revelation: A Radical Synthesis of Science and Spirituality.* New York, NY: SelectBooks, 2018.

Long, Jeffery, and Paul Perry, Paul. *Evidence of the Afterlife: The Science of Near-Death Experiences.* New York, NY: HarperCollins, 2010.

Malsburg, Von der. "Binding in Models of Perception and Brain Function," *Current Opinion in Neurobiology*, **5**, 1995.

Margenau, Henry. *The Miracle of Existence.* Woodbridge, CT: Ox Bow, 1984.

Moody Raymond A., and Paul Perry. *Coming Back: a Psychiatrist Explores Past-life Journeys.* New York, NY: Bantam, 1991.

Moody, Raymond A. *Life after Life: The Investigation of a Phenomenon—Survival of Bodily Death.* New York, NY: HarperCollins, 1975.

Moody, Raymond A. *Paranormal: My Life in Pursuit of the Afterlife.* New York, NY: HarperOne, 2012.

Moody, R. L. "Bodily Changes during Abreaction," *Lancet*, **1**, 1948.

Mott, Maryann. "Can Animals Sense Earthquakes?" *National Geographic News*, November 11, 2003.

Nagel, Thomas. *Mind and Cosmos: Why the Materialist neo-Darwinian Conception of Nature is Almost Certainly False.* Oxford: Oxford University Press, 2012.

Nahm, Michael, et al. "Terminal Lucidity: A Review and a Case Collection." *Archives of Gerontology and Geriatrics* 55, **1**, 2001.

Narasimhan, Ashok, and Menas Kafatos. "Wave Particle Duality, the Observer and Retrocausality," *Quantum Retrocausation III*, D. Sheehan (ed.) AIP Conference Proceedings, 1841:040004-1, 2016.

Narasimhan, Ashok, Deepak Chopra, and Menas Kafatos. "The Nature of the Heisenberg-von Neumann Cut: Enhanced Orthodox Interpretation of Quantum Mechanics," 2019.

Oldenburg, Dan. *Washington Post.* January 8, 2005.

Ouzts, Elizabeth. "In North Carolina, hog waste pollution a familiar result. Will things ever change?" Environmental Health News, Sept. 21, 2018.

Playfair, Guy. *Twin Telepathy.* Guildford, UK, White Crow, 2012.

Pollan, Michael. "The Intelligent Plant," *New Yorker Magazine*, Dec. 23, 2013.

Potts, Wayne. "The Chorus-Line Hypothesis of Manoeuvre in Avian Flocks," *Nature*, **309**, 1984.

Prophet Elizabeth C., and Erin L. Prophet. *Reincarnation: The Missing Link in Christianity.* Gardiner, MT: Summit Publications, 1997.

Puryear, Herbert B. *Why Jesus Taught Reincarnation.* Scottsdale, AZ: New Paradigm Press. 1993.

Puthoff, Harold, and Russell Targ. "A Perceptual Channel for Information Transfer over Kilometer Distances: Historical Perspective and Recent Research," *Proceeding of the Institute of Electrical and Electronic Engineers,* **64**, 1976.

Radin, Dean. *The Conscious Universe: The Scientific Truth of Psychic Phenomena.* New York, NY: HarperCollins, 1997.

Radin, Dean. *Entangled Minds: Extrasensory Experiences in a Quantum Reality.* New York, NY: Paraview, 2006.

Radin, Dean, et al. "Psychophysical Interactions with a Double-slit Interference Pattern," *Physics Essays* , **26**, no.4, 2013.

Rhine, Joseph, et al. *Extrasensory Perception after Sixty Years*. Boston, MA: Bruce Humphries, 1966.

Rhine, Joseph, et al. *Extrasensory Perception after Sixty Years: A Critical Appraisal Of The Research In Extra Sensory Perception*. New York, NY: Borodino Books, 2018. (Originally published in 1940).

Richheimer, Steven. *The Unity Principle: The Link between Science and Spirituality*. San German, PR: Innerworld Publications, 2013.

Richheimer, Steven. *The Nonlocal Universe: Why Science Validates the Spiritual Worldview*. San German, PR: Innerworld Publications, 2016.

Richheimer, Steven. *Reincarnation: Science of the Afterlife*. San German, PR: Innerworld Publications, 2019.

Salter, A. M. "Impact of Consumption of Animal Products on Cardiovascular Disease, Diabetes, and Cancer in Developed Countries." *Animal Frontiers*, **3**, 1, Jan. 2013.

Savage, Neil. "Seeking Materials to Send Unbreakable Codes," *Chemical & Engineering News*. September 11, 2017.

Schwartz, Stephan A. *Opening to the Infinite: The Art and Science of Nonlocal Awareness*. Budha, TX: Nemoseen Media, 2007.

Scoles, Sarah. "The God Mod: Psychedelic Drugs can Help Ease Depression, Anxiety, and Addiction, but why are those Results even better when Trip Takers have a Spiritual Encounter?" *Popular Science*, Fall, 2020.

Searle, John. *Rediscovery of the Mind*. Cambridge, MA: MIT Press, 1994.

Sheldrake, Rupert. *A New Science of Life: The Hypothesis of Morphic Resonance*. Los Angeles: J. P. Tarcher, 1981.

Sheldrake, Rupert. "Rat Learning and Morphic Resonance," 1995.

Sheldrake, Rupert. *Morphic Resonance: The Nature of Formative Causation*, Rochester, VT: Park Street Press, 1995. (rev. ed. of: *A New Science of Life*).

Sheldrake, Rupert. *Dogs That Know When Their Owners Are Coming Home: And Other Unexplained Powers of Animals*. New York, NY: Crown, 1999.

Sheldrake, Rupert. *The Sense of Being Stared At: And Other Aspects of the Extended Mind*. New York, NY: Random House, 2003.

Snyder, Timothy. *The Road to Unfreedom: Russia, Europe, America*. New York, NY: Tim Duggen Books, 2018.

Sperry, Roger. *The Encyclopaedia of Ignorance*. Eds. Duncan, R. and Weston-Smith, M. Oxford: Pergamon, 1977.

Stern, Jess. *Edgar Cayce the Sleeping Prophet: The Life, Prophecies, and*

Readings of America's Most Famous Mystic. New York, NY: Bantam, 1967.

Stevenson, Ian. *Telepathic Impressions: A Review and Report of 35 New Cases*. Charlottesville, VA: University of Virginia Press, 1970.

Stevenson, Ian. *Twenty Cases Suggestive of Reincarnation*, 2nd edition. Charlottesville, VA: University of Virginia Press, 1974.

Stevenson, Ian. *Reincarnation and Biology: A Contribution to the Etiology of Birthmarks and Birth Defects*. New York, NY: Praeger, 1997

Stevenson, Ian. *Children Who Remember Previous Lives: A Question of Reincarnation*. Jefferson, NC: McFarland, 2001.

Targ, Russell. *The Reality of ESP: A Physicist's Proof of Psychic Abilities*. Easton, IL: Quest, 2012.

Tart, Charles T. *The End of Materialism: How Evidence of the Paranormal is Bringing Science and Spirit Together*. Oakland, CA: New Harbinger, 2009.

The Dalai Lama. *The Universe in a Single Atom: The Convergence of Science and Spirituality*. New York, NY: Harmony, 2005.

Tiller, William. A., Walter Dibble, and Michael Kohane, Michael. *Conscious Acts of Creation: The Emergence of a New Physics*. Walnut Creek, CA: Pavior Publishing, 2001.

Treffert, Darold. "The Savant Syndrome: An Extraordinary Condition. A Synopsis: Past, Present, Future, Philosophical," *Transactions of the Royal Society*, **364**, 2009.

Tucker, Jim B. *Life Before Life: Children's Memories of Previous Lives*. New York, NY: St. Martin's Griffin, 2005.

Tucker, Jim B. *Return to Life: Extraordinary Cases of Children who Remember Past Lives*. New York, NY: St. Martin's Griffin, 2013.

Tyrrell, Toby. "It's a cosmic miracle that life on Earth's lasted this long." *Popular Science*, January 25, 2021.

Ullman, Robert, and Judyth Reichenberg-Ullman. *Mystics, Masters, Saints, and Sages: Stories of Enlightenment*. Berkeley, CA: Conari Press, 2001.

Watson, Lyall. *Lifetide: A Biology of the Unconscious*. London: Hodder and Stoughton, 1979.

Weiss, Brian L. *Many Lives, Many Masters: The True Story of a Prominent Psychiatrist, His Young Patient, and the Past-Life Therapy That Changed Both Their Lives*. New York, NY: Simon & Shuster, 1988.

Wheeler John, and Wojciech H. Zurek, eds. *Quantum Theory and Measurement*. Princeton, NJ: Princeton University Press, 1983.

Wheeler, John A. *At Home in the Universe*. Woodbury, NY: AIP Press, 1994.

Wheeler, John A. "From the Big Bang to the Big Crunch," *Cosmic Search Magazine*, **1,** no. 4, 1979.

Wheeler, John A. "Assessment of Everett's 'Relative State' Formulation of Quantum Theory." *Reviews of Modern Physics*, **29**, 1957.

Wiseman, Richard, Matthew D. Smith, and Julie Milton. "Can Animals Detect When Their Owners are Returning Home? An Experimental Test of the 'Psychic Pet' Phenomenon." *British Journal of Psychology*, **89**, no. 3, 1998.

ENDNOTES

1   Other terms for this worldview are *monistic idealism, holism, non-dualism, inseparability, fundamental oneness, advaita,* and *panentheism*. Although these terms may be used interchangeably, I prefer to use the word *spirituality* to describe this worldview. All these terms identify consciousness as the ultimate reality.

2   Of course, there is a full spectrum of ideas describing reality that fall between these two extremes, but they can be understood to belong to one category or the other.

3   One cloud was the results of the Michelson-Morley experiment, which did not detect the luminous aether, and the other was the failure of electromagnetic theory to explain the spectral distribution of black body radiation.

4   Relativity has spacetime as continuous, and interactions are local and deterministic, while quantum mechanics is discontinuous, indeterminate, and nonlocal. However, both point to wholeness as a basic feature of reality.

5   This is a body that absorbs all incident electromagnetic radiation, regardless of its frequency.

6   There is no question that describing quanta phenomena using a wave function has tremendous utility and produces extremely accurate predictions and results. The only problem is that a wave function cannot be detected by any known means or even visualized. This means that the wave function may be a subtle, hidden realm of reality responsible for the emergence of the physical world. We will cover this topic in detail in Chapter 6.

7   Steven Richheimer, *The Nonlocal Universe: Why Science Validates the Spiritual Worldview*, (San German, PR: InnerWorld, 2016), 29-30.

8   Nonlocality means that something is not localized in any particular place in space but can be thought of as being everywhere at the same time. It can also refer to instantaneous action at a distance and inseparability of particles having spacetime separation.

9   Theoretically, all the parts of the universe began fully connected in the inflation phase and like completely mixed milk and coffee everything was homogeneous. Hence, the theory of inflation explains why the temperatures and curvatures of different regions of space are so constant. The theory also allowed physicists to predict the minute differences in temperature of different regions of space in the primordial universe from quantum fluctuations during the inflationary era, and these predictions have been confirmed by observations of the remnants of the Big Bang—the cosmic microwave background radiation. Because of the very rapid growth of space during the inflation phase,
the universe expanded far beyond the size predicted if it were limited to the speed of light. Inflation also provides a mechanism whereby the universe could

initially emerge in a flat state following inflation.
10   Stephen Hawking and Roger Penrose, *The Nature of Space and Time* (Princeton: Princeton University Press, 1996), 89-90.
11   More precisely the Higgs field has a vacuum expectation value of 246.22 GeV (gigaelectron volts).
12   An analogy that is sometimes used is that of hitting a moving fly with a bullet shot from a gun from a mile away. Either the marksman was very good (God) or the gun fired an untold number of bullets.
13   Richard Feynman, *Feynman Lectures on Physics*, vol. 3, (Reading, MA: Pearson-Addison-Wesley, 1963), 1.
14   Yaakov Y. Fein, Philipp Geyer, Patrick Zwick, Filip Kiałka, Sebastian Pedalino, Marcel Mayor, Stefan Gerlich & Markus Arndt, "Quantum Superposition of Molecules beyond 25 kDa," *Nature Physics* volume **15**, (2019): 1242-5.
15   This is an example of the so-called "delayed-choice quantum-eraser experiment." See for example: Ashok. Narasimhan and Menas Kafatos, "Wave Particle Duality, the Observer and Retrocausality," *Quantum Retrocausation III*, Daniel Sheehan (edit) AIP Conference Proceedings, 1841:040004-1, 9 (2016).
16   The use of the word nonlocal means that a quantum is not localized in any particular place in space but can be thought of as being everywhere at the same time.
17   Neil Savage, "Seeking Materials to Send Unbreakable Codes," *Chemical & Engineering News*. September 11, 2017.
18   It can also be called quantum foundations.
19   One of the criticisms of the CI is that it requires a macroscopic entity such as a measuring device to interact with a microscopic quantum system. Collapse is then dependent on the interaction of a quantum system and classical observer. Where would you draw the line between quantum and classical? And if the wave function was for the universe who would be observer? It turns out this is not a valid criticism. According to CI, the quantum system is "questioned" by a measuring instrument. The quantum system then becomes part of a greater system that includes the quantum and macroscopic device. This is the same as saying the wave functions for the quantum and device become entangled.
20   More specifically decoherence is a loss of phase relationship between the quantum and the measuring device.
21   Sean Carroll, *The Big Picture: On the Origins of Life, Meaning, and the Universe Itself* (NY: Dutton, 2017), 169.
22   To quote Caltech physicist Sean Carroll: "…the wave function simply represents reality directly. Sean Carroll, *The Big Picture: On the Origins of Life, Meaning, and the Universe Itself* (NY: Dutton, 2017), 167.
23   Carroll writes, "There is only one wave function for the entire universe at once—what we call, with no hint of modesty, the wave function of the universe." Ibid. 168

24   John A. Wheeler, "Assessment of Everett's 'Relative State' Formulation of Quantum Theory", *Reviews of Modern Physics* 29, 1957, 462-465.

25   Paul Davies, *In Who We Live and Move and Have Our Being: Panentheistic Reflections on God's Presence in a Scientific World*. Editors: Philip Clayton and Arthur Peacocke (Grand Rapids, MI: Eerdmans Publishing, 2004).

26   David Bohm, *Wholeness and the Implicate Order* (NY: Routledge & Kegan Paul, 1983), 175.

27   Nick Herbert, *Quantum Reality: Beyond the New Physics* (Garden City, NY: Anchor Press/Doubleday, 1985), 18.

28   Larry Dossey, *Recovering the Soul: a Scientific and Spiritual Search* (NY: Bantam, 1989), 175.

29   John A. Wheeler, "From the Big Bang to the Big Crunch," *Cosmic Search Magazine*, 1 no. 4, 1979.

30   Kenneth Ford, "Update on John Archibald Wheeler" (PDF*). Princeton Physics News*, 2 (1) (Winter 2006).

31   John Gribbin, Mary Gribbin, and Jonathan Gribbin, *Q is for Quantum: An Encyclopedia of Particle Physics*. (NY: Simon and Schuster, 2000).

32   John A. Wheeler and Wojciech H. Zurek, eds. *Quantum Theory and Measurement* (Princeton, NJ: Princeton University Press, 1983), 192.

33   John A. Wheeler, *At Home in the Universe* (Woodbury, NY: AIP Press, 1994), 25.

34   Ibid. 45.

35   Interestingly the discovery of the Higgs particle in 2012 at CERN solved the puzzle of how particles acquire mass. However, this mass is acquired through an invisible substance now called the "Higgs field" that acts to impede a particle's movement when pushed through the field. Hence, the Higgs field is a modern version of an ether theory that says that the ether does interact with matter.

36   Stephen Hawking, ed. *A Stubbornly Persistent Illusion: The Essential Scientific Works of Albert Einstein* (Philadelphia, PA: Running Press, 2009), 56.

37   Gravitational Waves Detected 100 Years After Einstein's Prediction: https://www.ligo.caltech.edu/news/ligo20160211

38   One can argue that the degree to which today's religions deviate from spiritual ideology is simply a measure of how their followers have corrupted the original spiritual message of their founder.

39   One problem such an attempt to demonstrate "machine consciousness" might encounter is that a sufficiently advanced neural network with complexity similar to the human brain might conceivably be co-opted by a nonphysical entity (spirit?), which could take control of the computer and display elements of conscious awareness.

40   Carl Jung, *The Symbolic Life* (Princeton, NJ: Princeton University Press, 1976), para. 826.

41   Quoted in Amanda Gefter, *Trespassing on Einstein's Lawn* (NY: Bantam,

2017), 283.
42   Werner Heisenberg, *Physics and Philosophy: The Revolution in Modern Science* (Amherst, NY: Prometheus, 1958)129.
43   David Chalmers. *The Conscious Mind: In Search of a Fundamental Theory* (NY: Oxford University Press, 1996).
44   Christof Koch, "Proust among the Machines." *Scientific American*, Dec. 2019, 48.
45   John Searle, *Rediscovery of the Mind* (Cambridge, MA: MIT Press, 1994).
46   Von der Malsburg, "Binding in Models of Perception and Brain Function," *Current Opinion in Neurobiology*, **5**, (1995): 520-26.
47   Brian Greene, *Until the End of Time: Mind, Matter, and Our Search for Meaning in an Evolving Universe* (NY: Alfred Knopf, 2020), 158.
48   Carroll actually labels himself as a *poetic naturalist*.
49   Sean Carroll, *The Big Picture: On the Origins of Life, Meaning, and the Universe Itself* (NY: Dutton, 2017), 381.
50   Monistic idealism should not be confused with panpsychism. Panpsychism is the metaphysical view that everything—including a rock— *has* mind or consciousness. For monistic idealism, everything is *composed* of consciousness.
51   Richard Gerrig and Phillip Zimbardo, *Psychology and Life*, 20th Ed. (Essex England: Pearson Education, 2014).
52   Henry Margenau, *The Miracle of Existence* (Woodbridge, CT: Ox Bow, 1984), 96.
53   Edward F. Kelly and Emily W. Kelly, et al, *Irreducible Mind: Toward a Psychology for the 21st Century* (Lanham, MD: Rowman & Littlefield, 2007).
54   This means that neither the patient nor the persons administering the treatment are unaware whether they are receiving the experimental drug or treatment. This is because if the doctor or other healthcare professional knows whether they are administering a placebo or not, they might influence the patient outcome in subtle ways.
55   Edward F. Kelly and Emily W. Kelly, et al, *Irreducible Mind: Toward a Psychology for the 21st Century*, 129.
56   Lyall Watson, *Lifetide: A Biology of the Unconscious* (London: Hodder and Stoughton, 1979), 90.
57   R.L. Moody, "Bodily Changes during Abreaction," *Lancet*, 1, (1948): 964.
58   Edward F. Kelly and Emily W. Kelly, *Irreducible Mind: Toward a Psychology for the 21st Century*, 216.
59   L.K. Kothari, Arun Bordia, and O. P. Gupta, "The Yogic Claim of Voluntary Control over the Heart Beat: an Unusual Demonstration." *J. American Heart Assoc.*, **86**, no. 2, (1973), 284.
60   Edward F. Kelly and Emily W. Kelly, et al, *Irreducible Mind: To-ward a Psychology for the 21st Century*, 172-3
61   Ibid. 32.

62 Larry Dossey, *Healing Words: The Power of Prayer and the Practice of Medicine* (NY: HarperCollins, 1993).
63 Michael Nahm, Bruce Greyson, Emily Kelly, and Erlendur Haraldsson, "Terminal Lucidity: A Review and a Case Collection, *Archives of Gerontology and Geriatrics* 55, no. 1 (2001), 138-42. Available at: https://pubmed.ncbi.nlm.nih.gov/21764150/ [Accessed Jan. 2021].
64 Darold Treffert, "The Savant Syndrome: An Extraordinary Condition. A Synopsis: Past, Present, Future, Philosophical," *Transactions of the Royal Society*, 364 (2009), 1354.
65 Richard M. Bucke, *Cosmic Consciousness: A Revolution in the Study of the Human Mind* (NY: Dutton, 1969), 9-10
66 Robert Ullman and Judyth Reichenberg-Ullman, *Mystics, Masters, Saints, and Sages: Stories of Enlightenment* (Berkeley, CA: Conari Press, 2001), 37.
67 William James, *The Varieties of Religious Experience* (NY: Mentor, 1958), 295 (Original work published 1902).
68 Edward F. Kelly and Emily W. Kelly, et al, *Irreducible Mind: Toward a Psychology for the 21st Century*, 542-553.
69 Aldous Huxley, *Doors of Perception* (NY: Perennial Library, 1990) Originally published 1954.
70 In addition to their power for "opening the doors of perception," psychedelic drugs such as psilocybin have recently been found to help people cope with old and destructive mental patterns. Research from all over the world suggests that these drugs can break such patterns and help people fight addiction, alleviate depression, reduce existential fears, and improve relationships. For example see: Sarah Scoles, "The God Mod: Psychedelic Drugs can Help Ease Depression, Anxiety, and Addiction, but why are those Results even better when Trip Takers have a Spiritual Encounter?" *Popular Science*, Fall 2020, 52-9.
71 Robin Carhart-Harris, et al, "Neural Correlates of the Psychedelic State as Determined by fMRI studies with Psilocybin," *Proceeding of the National Academy of Sciences*, February 7, 2012, 109 (6) 2138-2143. Available from: https://www.pnas.org/content/109/6/2138 [Accessed Jan. 2021].
72 H. Hart, "ESP Projection: Spontaneous Cases and the Experimental Method," *Journal of the American Society for Psychical Research*, **48**, (1954): 121-46.
73 Edward F. Kelly and Emily W. Kelly, et al, *Irreducible Mind: Toward a Psychology for the 21st Century*, 389.
74 J. E. Owens, "Paranormal Reports from a Study of Near-death Experiences and a Case of an Unusual Near-death Vision." In L. Coly & J. McMahon (eds.), *Parapsychology and Thanatology*, (NY: Parapsychology Foundation, 1995), 149-167.
75 Sam Parnia, *Erasing Death: The Science that is Rewriting the Boundaries between Life and Death* (NY: HarperCollins, 2014).
76 Robert Bruce and Brian Mercer, *Mastering Astral Projection: 90-day Guide to Mastering Out-of-body Experience* (St. Paul, MN: Llewellyn, 2004).

77   Raymond A. Moody, *Life after Life: The Investigation of a Phenomenon—Survival of Bodily Death* (NY: HarperCollins, 1975).
78   https://www.nderf.org/NDERF/Research/number_nde_usa.htm.
79   Eben Alexander, *Proof of Heaven: A Neurosurgeon's Journey into the Afterlife* (NY: Simon & Schuster, 2012).
80   Jeffrey Long and Paul Perry, *Evidence of the Afterlife: The Science of Near-Death Experiences* (NY: HarperCollins, 2010).
81   See: https://iands.org/ndes/about-ndes/key-nde-facts21.html?start=3.
82   Nancy Clark, *Divine Moments: Ordinary People Having Spiritually Transformative Experiences* (Fairfield, IA: 1st World Publications, 2012).
83   Raymond A. Moody, *Paranormal: My Life in Pursuit of the Afterlife* (NY: HarperOne, 2012).
84   Dean Radin, *The Conscious Universe: The Scientific Truth of Psychic Phenomena* (NY: HarperCollins, 1997).
85   Dean Radin, *Entangled Minds: Extrasensory Experiences in a Quantum Reality* (NY: Paraview, 2006).
86   Russell Targ, *The Reality of ESP: A Physicist's Proof of Psychic Abilities* (Easton, IL: Quest, 2012).
87   Joseph Rhine, Gaither Pratt, Charles E. Stuart, Burke M. Smith and Joseph A. Greenwood, *Extrasensory Perception After Sixty Years: A Critical Appraisal Of The Research In Extra Sensory Perception* (NY: Holt, 1940).
88   Literally an analysis of analysis. A statistical study of a group of studies, each with its own statistical outcome.
89   Dean Radin, *The Conscious Universe: The Scientific Truth of Psychic Phenomena* (NY: HarperCollins, 1997), 79-80.
90   Ibid. 87-89.
91   Russell Targ, *The Reality of ESP: A Physicist's Proof of Psychic Abilities* (Wheaton, IL: Quest, 2012), 203-4.
92   Guy Playfair, *Twin Telepathy* (Guildford, UK, White Crow, 2012).
93   Dean Radin, *The Conscious Universe: The Scientific Truth of Psychic Phenomena*, 99-100.
94   Joseph Rhine, et al, *Extrasensory Perception after Sixty Years* (Boston: Bruce Humphries, 1966), 42.
95   Harold Puthoff and Russell Targ, "A Perceptual Channel for Information Transfer over Kilometer Distances: Historical Perspective and Recent Research," *Proceeding of the Institute of Electrical and Electronic Engineers,* **64** (1976).
96   Dean Radin, *The Conscious Universe: The Scientific Truth of Psychic Phenomena* (NY: HarperCollins, 1997), 105.
97   Robert G. Jahn, "The Persistent Paradox of Psychic Phenomena: An Engineering Perspective," *The Proceedings of the Institute of Electronic and Electrical Engineers,* **70**, no.2 (1982): 136-68.
98   Russell Targ, *The Reality of ESP: A Physicist's Proof of Psychic Abilities*

(Wheaton, IL: Quest, 2012), 12-13.
99 Ibid. 15
100 Ibid. 32
101 Russell Targ, *The Reality of ESP: A Physicist's Proof of Psychic Abilities* (Wheaton, IL: Quest, 2012), 50-1.
102 Ibid. 58-9
103 Ibid. 133-4.
104 Stephan A. Schwartz, *Opening to the Infinite: The Art and Science of Nonlocal Awareness*, (Budha, TX: Nemoseen Media, 2007).
105 R. G. Jahn, B. J. Dunne, & R. D. Nelson. "Engineering Anomalies Research," *Journal of Scientific Exploration*, **1** (1987): 21-50.
106 Dean Radin, *The Conscious Universe: The Scientific Truth of Psychic Phenomena* (NY: HarperCollins, 1997), 120.
107 Ibid. 126-130.
108 Ibid. 143.
109 Hardware RNGs are preferred and may employ radioactive decay or electronic noise. Noise spikes or decay signals are measured against an electronic oscillator such as a quartz crystal to produce thousands of spikes per second, which can be translated into a truly random sequence of zeros and ones that are accurately recorded by computer.
110 Dean Radin, *The Conscious Universe: The Scientific Truth of Psychic Phenomena* (NY: HarperCollins, 1997), 151.
111 B. J. Dunne and R. G. Jahn, "Experiments in Remote Human/Machine Interaction, *Journal of Statistics Edu.*, **6** (1992): 311-32.
112 Jahn, R. G., et al. "Correlations of Random Binary Sequences with Pre-Stated Operator Intention: A Review of a 12-Year Program Journal of Scientific Exploration," Vol. 11, No. 3, 1977. Available at: https://noosphere.princeton.edu/papers/pear/correlations.12yr.pdf [Accessed Jan. 2021].
113 Dean Radin, *Entangled Minds: Extrasensory Experiences in a Quantum Reality* (NY: Paraview, 2006), 195-202.
114 William A. Tiller., Walter Dibble, and Michael Kohane, *Conscious Acts of Creation: The Emergence of a New Physics* (Walnut Creek, CA: Pavior Publishing, 2001), 1-13.
115 Dean Radin, Leena Michel, James Johnston, and Arnaud Delorme, "Psychophysical Interactions with a Double-slit Interference Pattern," *Physics Essays* , **26**, no.4, (2013), 553-566.
116 Gabriel Guerrer, "Consciousness-related Interactions in a Double-slit Optical Interferometer, *Open Science Framework* (March 9, 2018).
117 Larry Dossey, *One Mind: How our Individual Mind is Part of a Greater Consciousness and Why it Matters*, (NY: Hay House, 2013), 20.
118 Ian Stevenson, *Telepathic Impressions: A Review and Report of 35 New Cases* (Charlottesville, VA: University of Virginia Press, 1970).

119 Larry Dossey, *One Mind: How our Individual Mind is Part of a Greater Consciousness and Why it Matters*, (NY: Hay House, 2013), 21

120 Sean Carrol, *Discover Magazine Blogs*, (2008). Originally accessed in June 2013 at: http://blogs.discovermagazine.com/cosmicvariance/2008/02/14/american-association-for-the-advancement-of-pseudoscience/#.XInSEqBKhPY. No longer available, but follow-up comment available at: https://www.preposterousuniverse.com/blog/page/57/?s=discover+magazine [Accessed Jan. 2021].

121 Now we might ask how it is possible for the mind to affect a desired probability event when normally all such events, for example, whether an electron will be observed with a clockwise or counterclockwise spin are random and thus unpredictable. One way to understand how this works is to consider the double-slit experiment in which we set up an electron detector at one of the slits to detect whether the electron goes through the slit. Doing this destroys the interference pattern and we get only a diffraction pattern. In other words, depending on how the experiment is set up in advance we can control *where* the electron goes on the detector. Hence, mental intent could cause a similar phenomenon to take place in the quantum mechanical brain and the effect could occur retroactively or before the cause (intention) just like in the delayed choice experiment.

122 Dean Radin, *The Conscious Universe: The Scientific Truth of Psychic Phenomena* (NY: HarperCollins, 1997), 81-2.

123 Ibid. 131-2.

124 Ibid. 100.

125 This effect has been studied and it has been shown that students that believe in psi (sheep) tend to score above chance while those that do not believe (goats) may actually score well below chance. The latter situation is known as "psi-missing" and the difference between the two groups can be highly significant. See: Charles Tart, *The End of Materialism: How Evidence of the Paranormal is Bringing Science and Spirit Together* (Oakland, CA, New Harbinger Publications, 2009), 137-8.

126 Elizabeth C. Prophet and Erin L. Prophet, *Reincarnation: The Missing Link in Christianity*, (Gardiner, MT: Summit Pubs. 1997).

127 Herbert B. Puryear, *Why Jesus Taught Reincarnation* (Scottsdale, AZ: New Paradigm Press, 1993).

128 Quincy Howe Jr., *Reincarnation for the Christian* (Wheaton, IL: Theosophical Pub. 1974).

129 Steven Richheimer, *Reincarnation: Science of the Afterlife*, (San German, PR: InnerWorld, 2019).

130 Ian Stevenson, *Children Who Remember Previous Lives: A Question of Reincarnation* (Jefferson, NC: McFarland, 2001), 30.

131 Ian Stevenson, *Twenty Cases Suggestive of Reincarnation*, 2nd edition (Charlottesville, VA: University of Virginia Press, 1974).

132 Ian Stevenson, *Children Who Remember Previous Lives: A Question of Reincarnation* (Jefferson, NC: McFarland, 2001).
133 Ian Stevenson, *Reincarnation and Biology: A Contribution to the Etiology of Birthmarks and Birth Defects* (NY: Praeger, 1997).
134 Jim. B. Tucker, *Life Before Life: Children's Memories of Previous Lives* (NY: St. Martin's Griffin, 2005).
135 Jim B. Tucker, *Return to Life: Extraordinary Cases of Children who Remember Past Lives* (NY: St. Martin's, 2013).
136 NBC Nightly News with Lester Holt on March 20, 2015. Available at: https://www.nbcnews.com/nightly-news/video/boy-remembers-amazing-details-of-past-life-as-hollywood-actor-416079939861 [Accessed Jan. 2021].
137 Raymond A. Moody and Paul Perry, *Coming Back: a Psychiatrist Explores Past-life Journeys* (NY: Bantam, 1991).
138 Brian L. Weiss, *Many Lives, Many Masters: The True Story of a Prominent Psychiatrist, His Young Patient, and the Past-Life Therapy That Changed Both Their Lives* (NY: Simon & Shuster, 1988).
139 Carol Bowman, *Children's Past Lives: How Past Life Memories Affect Your Child* (NY: Bantam, 1997).
140 Carol Bowman, *Return from Heaven* (NY: HarperCollins, 2001).
141 http://www.reincarnationforum.com.
142 Peter Fenwick and Elizabeth Fenwick, *Past Lives: An Investigation into Reincarnation Memories*, (NY: Berkley, 1999), 82-91.
143 Ian Stevenson, *Twenty Cases Suggestive of Reincarnation*, 2nd edition (Charlottesville, VA: University of Virginia Press, 1974).
144 Augustine, the Bishop of Hippo (354-430), converted to Christianity at the age of thirty-three. Prior to his baptism, he practiced Manichaeism, which taught a dualistic theology where there is a struggle between good (God) and an eternal evil power (the devil).
145 Sean Carroll, *The Big Picture: On the Origins of Life, Meaning, and the Universe Itself* (NY: Dutton, 2017), 166
146 Admittedly some of these are not "modern" physicists.
147 Notice that I am not saying that rocks possess or are conscious entities, but that they are ultimately *formed* from consciousness.
148 Known as *Prakriti* in Sanskrit.
149 This is called *sattvaguna* in Sanskrit, where *guna* means binding force.
150 This is known as *mahattattva* in Sanskrit, which means highest principle or element of reality.
151 This active principle of *Prakriti* is known as *rajoguna* in Sanskrit.
152 Called *tamoguna* in Sanskrit.
153 The Sanskrit term for the objective component of mind is *chitta*. Both cosmic mind and individual minds possess this component.
154 Since time begins with the creation of spacetime, which originates from

cosmic mind, the term "before" used here should be understood in terms of the sequence of events rather than in time.

155   One example of such interconnectedness is how rats that were trained to avoid one of two gangways that would give them an electric shock passed this knowledge onto subsequent generations. See: W. E. Agar, F.H Drummond, O.W. Tiegs, and M.M. Gunson, "Fourth (final) Report on a Test of McDougall's Lamarckian Experiment on the Training of Rats." *Journal of Experimental Biology,* **31**, (1954), 307-21.

156   Another example is the so-called "100th Monkey Phenomenon," which resulted from studies of an isolated group of Japanese monkeys; most of whom learned to wash the dirt from sweet potatoes, and this behavior was supposedly then learned by monkeys on a neighboring island.

157   Rupert Sheldrake uses the term "morphic resonance" to describe the process by which past forms and behaviors of organisms influence organisms in the present. For many other examples of such interconnectedness see his book: *Morphic Resonance: The Nature of Formative Causation,* Rochester, VT: Park Street Press, 1995. (rev. ed. of: *A New Science of Life*).

158   For example, physicist David Bohm calls the domain of the wave function the implicate order and directly identifies it as a mental realm.

159   I am referring here to the delayed choice quantum eraser experiment, in which knowledge or lack of knowledge of a photon's path determines whether they interfere or not, and such knowledge may be turned on or off well after the actual measurement was made. See also: https://en.wikipedia.org/wiki/Quantum_eraser_experiment.

160   Other physicists have taken this same position that a more rational interpretation of quantum mechanics entails the existence of a universal Observer that lies outside spacetime at an information level of existence. In other words, it is the availability of information to the observer, and not the presence of a human observer making measurements that is required for wave function collapse. See for example: Ashok Narasimhan, Deepak Chopra, and Menas Kafatos. "The Nature of the Heisenberg-von Neumann Cut: Enhanced Orthodox Interpretation of Quantum Mechanics," (2019). Available at: https://link.springer.com/article/10.1007/s41470-019-00048-x [Accessed Jan. 2021].

161   Some authors like to call cosmic mind the collective unconscious, one mind, or mental internet. These are merely other terms for this universal primordial mental state of creation.

162   These are termed *koshas* in Sanskrit. For more details about *the koshas* see my book, *The Unity Principle: The Link between Science and Spirituality* (San German, PR: InnerWorld, 2013).

163   In yoga, the *kosha* associated with the body is called the *annamaya kosha*.

164   Ferris Jabr, "How brainless slime molds redefine intelligence." *Nature,* 13 Nov. 2012. Available at: https://www.nature.com/news/

how-brainless-slime-molds-redefine-intelligence-1.11811 [Accessed Jan. 2021].
165   Michael Pollan, "The Intelligent Plant," *New Yorker Magazine*, Dec. 23, 2013.
166   Rupert Sheldrake, *A New Science of Life: The Hypothesis of Morphic Resonance* (Los Angeles: J. P. Tarcher, 1981).
167   I cover this topic in greater depth in my book, *The Unity Principle: The Link between Science and Spirituality* (San German, PR: InnerWorld, 2013).
168   Known as the *kamamaya kosha* in Sanskrit.
169   This layer of mind is called the *manomaya kosha* in Sanskrit. *Manomaya* literally means mental.
170   See: Shrii Shrii Anandamurthi, "Matter and Spirit," Published in *Ananda Marga Ideology and Way of Life in a Nutshell*, Part 5.
171   There is no separate existence of the unit's unconscious mind and cosmic mind. Any difference between the two is merely a theoretical proposition.
172   Carl Jung, *Man and His Symbols* (NY: Bantam, 1964).
173   Joseph Campbell, *The Hero with a Thousand Faces*, 3rd Ed. (Novato, CA: New World Library, 2008).
174   According to yogic philosophy when the *atman* is experienced in its pure egoless form, it is experienced as the *Paramatman*—Supreme Witnessing Entity of creation.
175   Noel Langley, *Edgar Cayce on Reincarnation*, (NY: Warner Books, 1967), 11.
176   Jess Stern, *Edgar Cayce the Sleeping Prophet: The Life, Prophecies, and Readings of America's Most Famous Mystic* (NY: Bantam, 1967).
177   Lacking any connection with physical reality, the bodiless mind may be considered to exist in a dormant or unconscious state like that of deep sleep. Therefore, ghosts cannot exist. However, if a person believes in the existence of ghosts and visits a house that is said to be haunted, fear may trigger a hallucination of a ghost. The concentration of mind caused by fear can leave an imprint in cosmic *objective mind* at that location. That in turn may trigger the subconscious minds of other people visiting the site to experience similar hallucinations. Therefore, ghost stories are typically propagated by fear and have no physical reality.
178   He called this cycle of death and rebirth *samsara*.
179   At that time, the sea was rich in potassium and low in sodium, but by the time that multicellular creatures developed about a billion years ago, the seas had changed from being potassium rich to sodium rich as more potassium containing compounds crystallized out and sodium dissolved. Hence, when circulatory systems developed they were more like seawater than cytoplasm, causing cells to adapt by developing a sodium-potassium pump that utilizes energy that could have been more productively utilized for other things.
180   Toby Tyrrell. "It's a cosmic miracle that life on Earth's lasted this long." *Popular Science*, January 25, 2021. Available at: https://www.popsci.com/story/science/earth-habitable-probability/ [Accessed Jan. 2021].

181 Amit Goswami, *Creative Evolution* (Wheaton, IL: Theosophical Publishing House, 2008), 15-20.
182 Michael J. Behe, *Darwin's Black Box: The Biochemical Challenge to Evolution* (NY: Free Press, 1996).
183 Amit Goswami, *Creative Evolution* (Wheaton, IL: Theosophical Publishing House, 2008), 178.
184 Thomas Nagel, *Mind and Cosmos: Why the Materialist neo-Darwinian Conception of Nature is Almost Certainly False* (Oxford: Oxford University Press, 2012).
185 Roger Sperry in R. Duncan and M. Weston-Smith, *The Encyclopaedia of Ignorance* (Oxford: Pergamon, 1977).
186 Robert Efron, "Biology without Consciousness—and its Consequences," In: *Logic, Laws and Life: Some Philosophical Complications,* ed. R. G. Colodny, (Pittsburgh: Pittsburgh University Press, 1977), 209.
187 Some viruses utilize RNA, but biologists do not classify them as living organisms since they can only reproduce by invading the cells of host organisms.
188 The theory also assumes that DNA holds within it the means of turning on such things as cell differentiation, organ design and protein synthesis in just the right order to insure the development and functioning of the intact organism.
189 The best scientific evidence for this comes primarily from studies of rats that were trained to run a maze, and this learned behavior was subsequently passed on to future generations. See Rupert Sheldrake. "Rat Learning and Morphic Resonance," 1995. Available at: https://www.sheldrake.org/essays/rat-learning-and-morphic-resonance [Accessed Jan. 2021].
190 Pam Belluck, "A Cat's 200-Mile Trek Home Leaves Scientists Guessing," *New York Times,* January 19, 2013.
191 Larry Dossey, *One Mind: How our Individual Mind is Part of a Greater Consciousness and Why it Matters,* (NY: Hay House, 2013), 54-5.
192 Wayne Potts, "The Chorus-Line Hypothesis of Manoeuvre in Avian Flocks," *Nature,* **309** (1984): 344-5
193 Rupert Sheldrake, *The Sense of Being Stared At: And Other Aspects of the Extended Mind,* (NY: Random House, 2003), 113
194 Maryann Mott, "Can Animals Sense Earthquakes?" *National Geographic News,* November 11, 2003.
195 Don Oldenburg, *Washington Post.* January 8, 2005.
196 Maryann Mott, *National Geographic News,* November 11, 2003.
197 Richard Wiseman, Matthew D. Smith, and Julie Milton, "Can Animals Detect When Their Owners are Returning Home? An Experimental Test of the 'Psychic Pet' Phenomenon." *British Journal of Psychology,* 89, no. 3 (1998): 453-62. Available at: https://www.researchgate.net/publication/13552100_Can_animals_detect_when_their_owners_are_returning_home_An_experimental_test_of_the_'psychic_pet'_phenomenon [Accessed Jan. 2012].

198 Rupert Sheldrake, *Dogs That Know When Their Owners Are Coming Home: And Other Unexplained Powers of Animals*, (NY: Crown, 1999).
199 Larry Dossey, *One Mind: How our Individual Mind is Part of a Greater Consciousness and Why it Matters*, (NY: Hay House, 2013).
200 Dean Radin, *Entangled Minds*: *Extrasensory Experiences in a Quantum Reality* (NY: Paraview, 2006).
201 John C. Eccles, *Facing Reality: Philosophical Adventures by a Brain Scientist* (NY: Springer-Verlag, 1970), 115.
202 Four-valued logic, unlike the two-valued logic that things are either true or false, includes both are true and false, and/or neither are true nor false. This can also be called quantum logic since like a two-valued qubit that can exist as a superposition of zero and one, when queried may become 00 (no, no); 01 (no, yes); 10 (yes, no); or 11 (yes, yes).
203 David Bohm, "On the Intuitive Understanding of Nonlocality as Implied by Quantum Theory," *Foundations of Physics*, **5** (1975).
204 Leon M. Lederman and Christopher T. Hill, *Quantum Physics for Poets* (Amherst, NY, Prometheus Books, 2011), 15.
205 Paul Levy, *The Quantum Revelation: A Radical Synthesis of Science and Spirituality* (NY: SelectBooks, 2018), 172.
206 The Dalai Lama, *The Universe in a Single Atom: The Convergence of Science and Spirituality*, (NY: Morgan Road, 2005), 66.
207 This is known as "promissory materialism." Alternate, non-materialistic explanations for unexplained phenomena are not needed since scientists in the future will eventually come up with a scientific explanation of such phenomena.
208 Perhaps the greatest failing of organized religions is the devastating effect of the hatred and fighting that has taken place throughout history among different religious factions. For more on the problems with organized religion see: Richard Dawkins, *The God Delusion* (NY: Mariner, 2008); Sam Harris, *Waking Up: A Guide to Spirituality without Religion* (NY: Simon & Shuster, 2014); Christopher Hitchens, *God is Not Great: How Religion Poisons Everything* (NY: Hachette, 2007).
209 Julianna M. Horowitz, et al. (2020). "Trends in Income and Wealth Inequality," Available at: https://www.pewsocialtrends.org/2020/01/09/trends-in-income-and-wealth-inequality/ [Accessed Jan. 2021].
210 For an excellent discussion of Russia's efforts to undermine the influence of the United States and European Union in order to spread the fascists ideology of Putin see: Timothy Snyder's *The Road to Unfreedom: Russia, Europe, America* (London: Vintage, 2018).
211 Kurt Anderson, *Evil Geniuses: The Unmasking of America: A Recent History* (NY: Random House, 2020), 369.
212 By 2016, estimates put the number of Americans addicted to opioids at 2 million Available at: https://www.asam.org/docs/default-source/advocacy/

opioid-addiction-disease-facts-figures.pdf.

213   A. M. Salter (2013). "Impact of Consumption of Animal Products on Cardiovascular Disease, Diabetes, and Cancer in Developed Countries." *Animal Frontiers*, 3, 1, Jan. 2013. Available at: https://academic.oup.com/af/article/3/1/20/4638623#198770575 [Accessed Jan. 2021].

214   Hana Kahleova, et al (2020). "Effect of a Low-Fat Vegan Diet on Body Weight, Insulin Sensitivity, Postprandial Metabolism, and Intramyocellular and Hepatocellular Lipid Levels in Overweight Adults." JAMA Network Open, 3 (11):e2025454. Available at: https://jamanetwork.com/journals/jamanetworkopen/fullarticle/2773291/ [Accessed Jan. 2021].

215   The best example of this is Vladimir Putin's Russia in which massive inequality is stabilized and reinforced by the politics of gloom and doom and the supplanting of policy by propaganda. For an excellent and extensive discussion of these problems, I would recommend the book by Timothy Snyder, *The Road to Unfreedom* (Vintage, 2019).

216   Charles T. Tart, *The End of Materialism: How Evidence of the Paranormal is Bringing Science and Spirit Together* (Oakland, CA: New Harbinger, 2009), 241-2.

217   Paul Levy, *Dispelling Wetiko: Breaking the Curse of Evil* (Berkeley, CA, North Atlantic Books, 2013).

218   Paul Levy, *The Quantum Revelation: A Radical Synthesis of Science and Spirituality*, 26.

219   David Bohm, *On Dialogue* (NY: Routledge, 2007), 58.

220   Paul Levy, *The Quantum Revelation: A Radical Synthesis of Science and Spirituality*, 216.

221   Katherine Bourzac, "Methane Emissions Reach Record Levels," *Chemical and Engineering News*, July 20, 2020, 20.

222   In the first two decades after its release, methane is 84 times more potent than carbon dioxide, but it doesn't linger as long in the atmosphere, and therefore overall it is considered to be only thirty times more devastating to the climate because of how effectively it absorbs heat than CO_2.

223   FAO, 2006. "Livestock's Long Shadow: Environmental Issues and Options," Food and Agriculture Organization of the United Nations.

224   Robert Goodland and J. Anhang, "Livestock and Climate Change: What if the Key Actors in Climate Change were Pigs, Chickens and Cows?" (2009). *Worldwatch*, November/December 2009, 10–19.

225   https://www.cowspiracy.com/facts

226   Elizabeth Ouzts, (2018). "In North Carolina, hog waste pollution a familiar result. Will things ever change?" *Environmental Health News*, Sept. 21, 2018. Available at: https://www.ehn.org/hurricane-florence-floods-north-carolina-hog-farms-2606610607.html [Accessed Jan. 2021].

227   Prout recognizes that certain kinds of firms must be very large in order to operate efficiently. These are known as natural monopolies, and should be

run by the government. An example is an electrical utility that benefits from having an ever-larger number of users because the additional cost per customer for maintaining its infrastructure declines as it grows.

228   Britta K. Hölzel, James Carmody, MarkVangel, Christina Congleton, Sita M. Yerramsetti, Tim Gard, and Sara W. Lazara, "Mindfulness Practice Leads to Increases in Regional Brain Gray Matter Density," *Psychiatry Research: Neuroimaging*, **191**, no. 1 (2011): 36–43.

229   Anandamurti's spiritual ideology is perhaps best described in his book, *Idea and Ideology*. However, I have referenced others books and sources of his. The great advantage of using the writings of Anandamurti is that by all indications he possessed the incredible ability to access at will the unlimited knowledge contained in cosmic mind.

230   The equivalent Sanskrit word is provided in italics.

231   https://www.britannica.com/science/intergalactic-medium

232   Bernard d'Espagnat, *In Search of Reality* (NY: Springer-Verlag, 1983).

233   Physicists call this "renormalization." For example, an electron has a surrounding field of energy that the mathematics indicates is infinite. To remove the infinite energy it is subtracted from both sides of the equation or renormalized. Recall that mass equates to energy and therefore the almost infinite energy of spacetime composed of tiny black holes is ignored in order to come up with a meaningful description of spacetime.

234   Its wavelength can be considered infinite or straight line. The I feeling (*mahattattva*) is the first quality imparted to cosmic consciousness and can be considered to have a finite, but very long wavelength.

235   This can occur with a high degree of probability but not certainty.

236   For an in-depth discussion of good, evil, happiness, and suffering I would suggest my book, *The Nonlocal Universe: Why Science Validates the Spiritual Worldview*.

INDEX

A

abiogenesis 192, 193, 194
absolute reality 76, 87, 88, 250
acupuncture 176, 179, 218, 227
Adams, Robert 98
aerial factor 156, 245
Age of Mammals 195
akasha 170, 243
Alexander, Eben. 107, 110, 260, 273
Amazon River 228
amnesia 105, 109, 181
amygdala 236
ananda 101, 259
Ananda Marga (Path of Bliss) vi, vii, 260, 278
Anandamurti, Shrii Shrii v, vi, vii, xi, xii, 151, 241, 282
anatomical convergence 83
Andromeda Galaxy 60
animal agriculture 222, 233
animal instincts 178, 201, 202, 206
anomalies xi, 4, 21, 85, 141, 145, 146, 203, 208, 239
anthropic principle 26, 194
archetypes 168, 177, 178, 202, 214, 254
Aristotle 181, 215
arrow of time 22, 27, 53, 56, 65, 163, 164, 171
artificial intelligence (AI) 37, 68, 91

Aspect, Alain 35
astral projection 105, 183
atman 178, 278
Augustine (Bishop of Hippo) 138, 139, 276
autism 137
avidya 256
AWARE studies 105
Ayurveda 176, 227

B

Baha'u'llah 98
Bauer, Edmond 144
Behe, Michael 196, 260, 279
Bell, John 34, 35, 36, 144
Bell's theorem 35, 36
bhakti 237, 258, 259
bhakti yoga 258
Bible 185
Big Bang 20, 21, 22, 23, 24, 25, 26, 27, 49, 56, 61, 65, 83, 124, 153, 154, 166, 172, 241, 242, 244, 245, 267, 268, 270
biofeedback 93, 227
birth defects 94, 131, 132, 139, 140, 186
birthmarks 94, 131, 132, 139, 140, 145
bits 10, 36, 120, 121
black-body radiation 141
black hole 59, 63, 247
block time 62, 251
blood clotting system 196
Bobbie, the dog 204, 205

bodiless mind 138, 190, 252, 278
Bohm, David 45, 46, 76, 144, 213, 225, 247, 249, 260, 270, 277, 280, 281
Bohmian mechanics 45, 46
Bohr, Neils 30, 34, 38, 40, 47, 54, 144, 167, 250
bowerbird 201
Bowman, Carol 135, 260, 276
Brahma 153, 207, 242, 243
Bucke, Richard M. 97, 260, 272
Buddha 98, 191, 255
Buddhism 130

C

Cambrian Period 195
Campbell, Joseph 178, 260, 278
capitalism 223, 235
Capra, Fritjof 144
cardiac arrest 11, 79, 89, 92, 105, 106, 108, 138, 183
Carrol, Sean 275
Catholic 1
Cayce, Edgar 185, 186, 263, 265, 278
C. elegans (roundworm) 78, 201
Cenozoic era 195
Chalmers, David 81, 86, 261, 271
Charon, Jean 144
Chaser (the dog) 159
chiropractic 218, 227
Chopra, Deepak 264, 277
Christ 93
Christianity 130, 138, 264, 275, 276
Christian Science 179, 227
clairvoyance 104, 113, 114, 116
Clark, Nancy 110, 261, 273
classical physics 15, 30, 60
climate change 233
collapse of the wave function 16, 19, 27, 40, 41, 53, 56, 165
collective unconscious 127, 177, 178, 277
communism 235
complementarity 40, 54, 249, 250
conscious mind 176, 177, 183, 253, 256
consumerism 217, 224, 226, 227
cooperatives 235
Copenhagen Interpretation (CI) 38
cosmic consciousness vii, xii, 2, 4, 49, 88, 96, 101, 102, 138, 150, 153, 154, 158, 160, 161, 164, 166, 172, 187, 189, 190, 191, 198, 207, 231, 232, 233, 235, 237, 239, 240, 242, 243, 244, 247, 249, 250, 254, 255, 256, 257, 258, 259, 282
Cosmic Entity xii, 153, 237, 249, 257
cosmic microwave background radiation 268
cosmic mind x, 3, 4, 49, 56, 76, 100, 127, 146, 154, 156, 157, 158, 159, 160, 164, 165, 166, 167, 168, 169, 170, 171, 176, 177, 178, 181, 183, 184, 185, 186, 194, 198, 199, 202, 204, 205, 206, 207, 208, 215, 232, 242, 243, 244, 245, 246, 247, 248, 249, 251, 252, 253, 254, 257, 276, 277, 278, 282
COVID-19 213, 223, 234

creative principle (CP) 4, 150, 153, 154, 170, 181, 207, 242, 243, 244
Creed of Scientific Materialism 10
cryptomnesia 139
cyanobacteria 157
cyclotron 58

D

dark energy 245, 246
dark matter 246
Darwin, Charles 8, 28, 194, 195, 196, 260, 279
Darwinism 194, 195, 196, 197, 198, 262
Dawkins, Richard 195, 261, 280
Dead Sea Scrolls 186
déjà vu 136
delayed choice experiment 49, 275
democracy 223, 224
Descartes, Rene 7, 75, 79, 87
d'Espagnat, Bernard 247, 261, 282
determinism 10, 18, 19, 26, 27, 41, 43, 200, 213, 214, 250
devayonis (luminous bodies) 254
devotion 254, 258
diffraction pattern 32, 275
dissociative identity disorder (DID) 136, 180
Division of Perceptual Studies (University of Virginia School of Medicine) 91, 123, 131, 133, 134
DMT 109
DNA x, 24, 121, 175, 176, 192, 193, 194, 196, 200, 201, 202, 279
dogma 214, 216, 229, 232, 238
domain of the wave function 3, 47, 53, 55, 56, 65, 125, 142, 144, 146, 150, 163, 164, 165, 169, 171, 184, 247, 249, 251, 277
Dossey, Larry 94, 123, 207, 261, 270, 272, 274, 275, 279, 280
double-slit experiment 30, 32, 33, 39, 49, 50, 122, 167, 275
dreams 103, 109, 133, 190
Druze 130
dualism 7, 87, 88, 139, 268
Dyson, Freeman 144

E

Earth 1, 12, 20, 23, 24, 25, 44, 50, 57, 59, 60, 61, 62, 63, 64, 106, 121, 145, 157, 160, 169, 192, 194, 197, 199, 203, 220, 225, 228, 232, 238, 254, 266, 278
Eastern religions 130, 191
ecstasy 11, 178, 191, 240, 258, 259
Eddington, Arthur Stanley 59
EEG 77, 108, 109
ego 88, 102, 159, 160, 161, 166, 181, 187, 189, 233, 235, 237, 255, 256, 257, 258
Einstein, Albert 3, 9, 15, 18, 22, 30, 34, 35, 45, 47, 52, 57, 58, 59, 62, 64, 65, 101, 125, 137, 166, 167, 168, 170, 251, 261, 262, 270
electroencephalograph (EEG) 77
electromagnetic force 23, 24
Empedocles 215
endorphins 109
engrams 90

enlightenment 162, 240, 258
entanglement xii, 18, 33, 35, 36, 37, 39, 125, 143, 146, 169, 216, 248, 251
entropy 21, 22, 27, 171, 172, 244
Eocene Epoch 195
epigenetics 202
epiphenomenalism 10, 84, 111
Er 106
espionage 116
ESP, see extrasensory perception 3, 4, 12, 95, 104, 113, 117, 139, 140, 218, 266, 272, 273, 274
eternal damnation 139
eternal now 64, 101, 251
euglenas 174
evil 1, 139, 256, 276, 282
evolution of species 194, 195, 198
explicate order 46, 249
extrasensory perception (ESP) 113

F

Feynman, Richard 31, 261, 269
fine-tuning 23, 24, 25, 26, 194, 243
fish-skin disease (ichthyosis) 93
four-dimensional sight 62
free will 11, 63, 83, 84, 143, 160, 168, 189, 198, 256
Friedman, Norman 144
fundamental factors 242, 243, 246, 252, 254

G

Galileo 1, 137, 220
ganzfeld telepathy experiments 126
general theory of relativity 58
genetic determinism 200
genius 94, 130, 136, 137, 139
ghosts 252, 253, 254, 278
Global Consciousness Project 121
global neuronal workspace (GNW) theory of consciousness 81
Gnostic Christians 131
God x, xii, 1, 2, 7, 11, 22, 26, 96, 101, 106, 107, 110, 130, 138, 161, 162, 172, 189, 207, 217, 219, 229, 240, 242, 256, 257, 258, 259, 261, 262, 265, 269, 270, 272, 276, 280
Golden Rule 188
good xi, 3, 11, 12, 17, 97, 104, 110, 120, 125, 130, 138, 139, 175, 179, 188, 189, 191, 198, 213, 233, 235, 256, 269, 276, 282
Gorilla Sign Language 159
Goswami, Amit 144, 262, 279
GPS system 64
gravitational constant 247
gravitational force 155, 241, 245
gravitational waves 64
gravity 8, 23, 24, 28, 58, 59, 63, 65, 155, 168, 241, 245
Greene, Brian 84, 262, 271
guna 207, 276
guru vi, 97, 255

H

Haicheng, China 206
hallucination 103, 154, 252, 253, 278

Hammons, Ryan 133
hard problem 80, 81, 88, 180, 181, 212
Hawking, Stephen 23, 262, 269, 270
Hearst, Patricia 117
heat death of the universe 245

I

idealism xii, 87, 268, 271
ideology 4, 87, 95, 128, 141, 141, 147
IIED (see intention imprinted electrical devices)
immortality..215, 232
immune system 12, 92, 197, 200
implicate order 46, 247, 249
income inequality 221, 222, 224
individualism 227, 235
inflation (cosmic) 21, 25, 26, 243, 244, 268
integrated information theory (IIT) 81
intention imprinted electrical devices (IIED) 121, 127
intentionality 3, 88, 91, 180, 181
interference pattern 31, 32, 33, 122, 275
intergalactic hydrogen 245
irreducibly complex 196
Islam 130
Ismaili 130

J

Jainism 130
Jeans, James 144
Jesus 98, 131, 138
jinana 258

Judaism 98, 130
Jung, Carl 74, 177, 178, 236
Jupiter 117, 156, 220

K

Kafatos, Menas 144, 264, 269, 277
karma 129, 130, 138, 186, 187, 188, 189, 191, 224, 233, 257
Kelly, Edward F. 91, 263, 271, 272
ketamine 109
Koko (the gorilla) 159

L

Lamarckism 201, 202
Lao Tzu 98
law of conservation of energy 19, 20
law of karma 130, 186, 187, 188, 189, 191, 224, 233
layers of mind 179
liberal education 223
liberation 162, 191, 229, 254, 258
LIGO observatories 64
liila 207
limitlessness 208, 232, 238, 239, 257
lipid bilayer 193
London, Fritz 144, 202, 266, 271, 280
Long, Jeffery 108, 263, 273, 281
LSD 99, 103, 109
luminous bodies 254
luminous factor 155, 156

M

macrocosm 54, 177, 239, 248, 250
madhuvidya 257
magnetoencephalograph (MEG) 77
mantra 218
many-worlds interpretation (MWI) 41, 45, 124, 142, 246
Margenau, Henry 91, 144, 263, 271
Mars 62, 156, 157
Martyn, Marty 133, 134
materialism x, 1, 2, 3, 4, 7, 8, 9, 10, 12, 14, 17, 18, 19, 25, 26, 40, 55, 68, 69, 84, 87, 95, 124, 128, 141, 142, 144, 145, 146, 147, 171, 173, 180, 181, 211, 214, 216, 217, 219, 220, 224, 225, 226, 227, 228, 229, 230, 238, 280
materialist ideology 226
materialist worldview xi, 4, 7, 8, 10, 23, 26, 41, 66, 68, 71, 76, 83, 84, 140, 141, 144, 145, 146, 163, 171, 182, 208, 214, 216, 217, 219, 223, 224, 238
McMoneagle, Joe 117
meditation v, vi, vii, 93, 96, 100, 105, 190, 191, 227, 234, 236, 237, 257
memory 11, 73, 78, 79, 81, 88, 89, 90, 94, 95, 106, 108, 109, 133, 136, 137, 139, 159, 174, 175, 177, 198, 201
Mercury 156
mescaline 99
meta-analysis 115, 120, 126
metaphysics x, 38, 216, 219
meteorites 220
methylenedioxymethamphet-amine (MDMA) 100
Michelson, Albert 8, 57, 268
microcosm 54, 169, 177, 239, 248, 250
midlife crisis 236
Milky Way 62, 245
mind-body unity 92, 95, 173
mindfulness meditation 236
Mind of God 207
mind virus 225, 226
Mohammed 98
moksha 162
momentum 17, 30, 34, 54, 188, 189, 250
monarch butterflies 203
monistic idealism 87, 268, 271
Moody, Raymond A. 107, 110, 111, 134, 263, 264, 271, 273, 276
Moon 167
morality 11, 235
Morley, Edward 57, 268
Moses 98
Mozart 137
mukti 162
multicellular organisms 158, 159, 174
multiverse 25, 26, 243, 244
Myers, F.W.H. 92, 114
mystical experience 96, 97, 98, 99, 100, 102, 178, 183, 251

N

Nagel, Thomas 197, 198, 264, 279
nationalism 222, 223
natural selection 195, 198
naturopathy 227
near-death experience (NDE)

103, 105, 107, 110, 111
Nelson, Roger 121, 131, 263, 274
neo-Darwinism 194, 195, 196, 197, 198
Neoplatonists 131
Neptune 156
network model (memory) 78, 176
Neumann, John von 144, 264, 277
neuron 19, 77, 78
neuroscience 26, 68, 81, 82, 83, 85, 95, 107, 111, 173, 180, 181, 212
New Age 212, 239
Newton, Isaac 8, 22, 60, 137, 187
Nirguna Brahma 207, 242, 243
nirvana 162, 258
nirvikalpa samadhi 94
nocebo 92
noncerebral memories 190
nonlocality 18, 31, 32, 33, 35, 45, 55, 141, 146, 184
nonphysical x, xi, 3, 11, 19, 20, 27, 39, 40, 41, 43, 48, 56, 68, 72, 84, 92, 106, 124, 129, 130, 138, 139, 142, 146, 165, 168, 169, 174, 175, 180, 182, 184, 216, 219, 227, 252, 270
Nostradamus 185
nucleus of creation 242

O

objective mind (OM) 154, 158, 162, 170, 175, 176, 188, 190, 207, 243, 278

occult powers 255
Old Testament 139
One Mind 91, 123, 207, 261, 274, 275, 279, 280
ontology 2, 9, 14, 17, 43, 45, 46, 49, 68, 69, 87, 142, 145, 150, 173, 182, 219, 229, 237
Oracle of Delphi 185
Origen 131
original sin 138
Oscar (the cat) 137, 206
out-of-body experience (OBE) 103, 105

P

panaceas 215
pandemics 224, 234
panentheism 268
paradigm shift 217, 218
paramecium 174
paranormal beings 252
paranormal phenomena 123, 216, 239, 242
parapsychology 113, 124, 146
Parnia, Sam 105, 272
past-life regression 135
Pauli, Wolfgang 144
Pavšič, Matej 144
PEAR, Princeton Anomalies Research Laboratory 118, 120, 121
Peek, Kim 137
Penrose, Roger 144, 262, 269
phobias 134, 135, 191
photoelectric effect 30
photon xii, 31, 32, 36, 37, 50, 60, 167, 169, 170, 277
photosynthesis 157, 174
physicalism 2, 81

placebo 3, 92, 179, 271
Planck constant 247
Planck epoch 242
Planck, Max 9, 14, 30, 144, 242, 247
planets 21, 23, 24, 55, 98, 156, 194, 220, 254, 255
plasma 155
Plato 66, 106, 168
polarization 35, 169
pollution 202, 228, 264, 281
Prabhat Samgiita vii
Prakriti 207, 242, 276
Pratt, J.G. 116, 273
prayer 94, 96, 227
precognition 113, 118, 119, 122, 170, 184, 248, 252
presentiment 119, 126
Price, Pat 117
Prigogine, Ilya 144
Princess Diana 121
Principe Island 59
probability 12, 15, 16, 17, 20, 25, 30, 34, 39, 40, 47, 52, 56, 69, 118, 165, 167, 193, 194, 196, 275, 278, 282
Project Deep Quest 118
Prophet, Elizabeth C. 264, 265, 275, 278
proteins 24, 176, 192, 193, 194, 197, 200, 201
protoplanet 155
Prout (Progressive utilization theory) 234, 235, 281
provincialism 223
pseudoscience 13, 124, 143, 146, 216, 218, 275
psilocybin 260, 272
psi phenomena 95, 113, 122, 123, 125, 126, 127, 146, 182, 184, 216, 220, 251, 252
psychic abilities 206
psychic body 105, 174, 175, 176, 177, 179, 183, 190, 252
psychic phenomena 12, 113, 128
psychic powers 105
psychokinesis 113, 120, 184, 252
psychoneuroimmunology 92
Purushottama 242
Puryear, Herbert 264, 275

Q

qualia 3, 12, 80, 86, 88, 95, 180, 227
quantum Bayesianism 46
quantum black holes 247
quantum communication 37
quantum computers 36, 37
quantum field theory 65
quantum mechanics 2, 3, 9, 14, 15, 16, 17, 18, 19, 20, 29, 30, 31, 34, 35, 38, 40, 41, 43, 44, 45, 46, 48, 50, 53, 54, 65, 68, 84, 124, 141, 142, 143, 144, 146, 164, 166, 168, 171, 172, 184, 199, 213, 214, 215, 220, 239, 241, 244, 246, 247, 248, 249, 250, 268, 277
quantum physics x, xii, 2, 9, 10, 17, 19, 26, 30, 36, 65, 76, 166, 171, 213, 214, 216, 226, 242, 249
Quantum Physics-Induced Trauma 214
quantum teleportation 37
quantum wave function 3, 15, 16, 51, 55
quasar 50

qubits 36, 37

R

Radin, Dean 120, 122, 207, 264, 273, 274, 275, 280
Rae, Alastair 144
reactive momenta 188, 189, 190, 191, 252, 255, 256, 258
reciprocal apparitions 104
reductionism 2, 55, 92, 146, 212, 213
reincarnation 3, 12, 95, 123, 130, 131, 132, 134, 135, 136, 138, 139, 140, 161, 182, 186, 187, 188, 190, 191, 218
Reincarnation Forum 135
relativity theory 9, 26, 53, 60, 62, 66, 68, 76, 141, 168, 169, 170, 171, 172, 248
religion 1, 7, 68, 97, 108, 124, 216, 217, 228, 229, 239, 280
religious fundamentalism 229
remote viewing 104, 113, 116, 117, 118, 122, 126, 184, 252
Republic (of Plato) 106
resurrection 138
Rhine, Joseph 114, 116, 265, 273
RNA 192, 193, 194, 279

S

Sagan, Carl 139
sages xi, 138, 150, 170, 173, 184, 207, 258
Saguna Brahma 207, 243
salvation 1, 138, 139, 162, 258
samadhi 94, 162, 258
samskara (see also reactive momenta) 188, 254
Sanskrit xii, 170, 188, 207, 242, 254, 256, 257, 258, 276, 277, 278, 282
Sarkar, P.R. vi, xi, 232, 234, 235
satori 162, 258
Saturn 156
Satyamurti, yogi 94
savant 94, 137
Schrödinger, Erwin 15, 16, 30, 38, 40, 42, 43, 50, 53, 74, 84, 144, 166
Schrödinger's Cat 42
Schrödinger wave equation 16, 42
Schwartz, Stephan 118, 265, 274
Science Applications International (SAIC) 117
scientism 124, 127, 218, 225
Searle, John 82, 265, 271
seeker 256, 257
seer 257
self-awareness 3, 68, 73, 75, 81, 83, 84, 88, 89, 90, 111, 129, 159, 176, 180, 181, 198, 236, 237, 239
selfless service 191, 233, 237, 257, 259
self-realization vii, 161, 162, 181, 189, 191, 235, 237, 254, 256, 258
sense organ 71, 175, 176
Sheldrake, Rupert 175, 205, 206, 265, 277, 278, 279, 280
Shiva 98
siddhis 105, 255
Sikhism 130
sin 138
skepticism 127, 135, 218
skin-writing 93
slime mold (Physarum polyceph-

alum) 174
socialism 235
soul 101, 106, 107, 110, 129, 138, 178, 235, 239, 254
spacetime 3, 22, 27, 28, 34, 35, 39, 40, 46, 57, 58, 59, 60, 61, 62, 63, 64, 65, 66, 101, 102, 113, 154, 155, 162, 164, 166, 167, 170, 171, 216, 242, 243, 244, 245, 246, 247, 248, 249, 251, 268, 276, 277, 282
spatial nonlocality 55
special theory of relativity 57, 58
spiritual ideology v, xi, xii, 2, 146, 151, 153, 154, 164, 166, 171, 176, 181, 187, 190, 191, 207, 218, 232, 236, 237, 238, 239, 241, 242, 244, 246, 249, 252, 256, 270, 282
spirituality vii, x, xii, 2, 3, 68, 69, 86, 143, 146, 171, 172, 174, 181, 197, 202, 207, 214, 215, 217, 228, 229, 248, 249, 258, 268
spiritual philosophy 153, 235
spiritual union vii, 130, 162, 187, 256
spiritual worldview xii, 4, 50, 65, 68, 76, 88, 102, 140, 143, 206, 232, 238, 239
SRI (Stanford Research Institute) 116, 117
Standard Model 241
Stapp, Henry 144
stars 21, 23, 24, 25, 27, 55, 59, 98, 155, 203, 220, 245, 254, 255
Star Trek 37
static creative principle 170, 244
Stevenson, Ian 123, 131, 132, 133, 135, 136, 145, 266, 274, 275, 276
St. Francis of Assisi 93
stigmata 3, 180
strong nuclear force 23
subconscious mind 109, 176, 177, 178, 179, 183, 253
subjective experiences x, 3, 73, 80, 81, 87, 180
suffering 93, 94, 104, 108, 130, 136, 139, 185, 188, 189, 191, 223, 225, 233, 256, 258, 282
Sufism 98
sun 1, 12, 22, 59, 60, 62, 65, 155, 156, 157, 192, 203, 258
supercomputers 36
superimplicate order 46, 249
supernatural 26, 126
supernova 60, 155
superposition 15, 16, 19, 27, 36, 37, 39, 40, 41, 42, 43, 44, 53, 166, 199, 252, 280
Swann, Ingo 116, 117

T

Talbot, Michael 253
tamoguna 276
Tantra vii
Taoism 130
tarantula hawk (Pepsis marginata) 197
Targ, Russell 115, 116, 117, 264, 266, 273, 274
telepathic impressions 123
telepathy 113, 114, 115, 116, 122, 126, 184, 252
temporal nonlocality 32, 33
terminal lucidity 94
The Dalai Lama 216, 266, 280

Tiller, William A. 121, 266, 274
time travel 169, 248
Tlingit Indians 130, 132
tsunami 205
Tucker, Jim B. 133, 134, 145, 266, 276

U

uncertainty principle 17, 19, 29, 54, 250
unhappiness 11, 188, 236, 258
unit consciousness 75, 158, 161, 178, 254
United Nations Food and Agriculture Organization 228
unit mind 76, 129, 130, 138, 158, 159, 161, 174, 175, 176, 177, 181, 183, 190, 248
unity of self-awareness 83, 89
universalism 223, 232
Upanishad 101
Uranus 156

V

Vedantism 98
Venus 156

W

wave function xii, 3, 15, 16, 17, 18, 19, 20, 27, 30, 33, 35, 38, 39, 40, 41, 43, 44, 45, 46, 47, 49, 51, 52, 53, 54, 55, 56, 65, 68, 122, 124, 125, 142, 143, 144, 146, 150, 163, 164, 165, 166, 167, 168, 169, 171, 184, 215, 216, 241, 247, 248, 249, 251, 252, 268, 269, 277
weak force 23, 24
weakly interacting massive particles (WIMPS) 246
Weiss, Brian 135, 266, 276
Wetiko 225, 263, 281
Wheeler, John A. 45, 47, 48, 49, 74, 144, 168, 246, 261, 266, 267, 270
Wigner, Eugene 144

X

Xenoglossy 135

Y

Y2K 121
Yala National Park 206

Z

Zeilinger, Anton 144

www.ingramcontent.com/pod-product-compliance
Lightning Source LLC
Chambersburg PA
CBHW021140080526
44588CB00008B/141